SCIENCE AND OCCULTISM

SCIENCE AND OCCULTISM

I. K. TAIMNI

THE THEOSOPHICAL PUBLISHING HOUSE
Adyar, Madras 600 020, India
Wheaton, Ill., USA • London, England

First Edition 1974
First Reprint 1987

ISBN

Printed at the Vasanta Press
The Theosophical Society,
Adyar, Madras 600 020, India

PREFACE

SINCE man has developed the capacity to wonder and think about the problems of human life and the universe, thoughtful individuals have been wondering and enquiring about the nature of the three basic realities of existence—Man, God and the universe. As far as the outer world is concerned this kind of enquiry has been confined within two well-marked channels. One channel represents the flow of thought derived from the revelations made in various religions which have appeared in the world from time to time. This is called theology. The other channel represents the flow of thought derived from speculation. This is called speculative philosophy.

The main current of thought with regard to the deeper problems of life had been flowing along these two channels until Science came upon the scene and adopted a new approach to these problems. This approach is based upon experimentation and investigation of all phenomena of Nature by means of objective and well-defined methods, and it has given rise to a new current of thought, partly experimental and partly speculative. This is called Science.

While the first two methods of enquiry have come down to us from times immemorial the new approach is of comparative recent origin because Science began its career in the present world only about three hundred years ago, as far as our present historical records go.

All the three branches of knowledge mentioned above—Religion, Philosophy and Science—take it as an axiomatic truth that there is an ultimate reality underlying the universe and all facts of existence. Conceptions with regard to the nature of this reality may differ, methods of finding that reality may vary, but there is no difference of opinion as far as the existence of such a reality is concerned. Most thoughtful people will concede the validity of the above statement as far as Religion and Philosophy are concerned but there may be some who doubt whether Science proceeds on the assumption that there is such a reality underlying the phenomenal world which is the object of its investigation. The following extracts from *The Universe and Dr. Einstein* will show that such is actually the case:

'The philosopher and mystic, *as well as the scientist*, have always sought through their various disciplines of introspection to arrive at a knowledge of the ultimate immutable essence that undergirds the mutable illusory world.'

'My religion', says Einstein 'consists of a humble admiration of the illimitable superior spirit who reveals himself in the slight details we are able to perceive with our frail and feeble minds. That deeply emotional conviction of the presence of a superior reasoning power which is revealed in the incomprehensible universe, forms my idea of God.'

The discipline of each of these fundamental branches of knowledge tries to tackle the problem of knowing this underlying reality in its own characteristic way. Let us briefly review these methods and point out their advantages and disadvantages before we deal with the Occult method of tackling the problem of knowing this Reality.

The inductive method of Science depends upon experimentation and the accumulation of accurate data derived from experiments performed to solve a particular problem. When sufficient data is available an effort is made to discover, if possible, the law underlying the phenomenon which is being investigated and to formulate a hypothesis which can account satisfactorily for the facts which have been observed. In the light of this hypothesis more experiments are undertaken to see whether the hypothesis remains valid under the new conditions. If the new facts discovered in this manner do not support the hypothesis it is either suitably modified or abandoned altogether and a new hypothesis substituted in its place. This process of experimentation and formulation of hypothesis is continued until a fairly satisfactory hypothesis which accounts for all the observed facts to a reasonable extent is arrived at. A comprehensive hypothesis which accounts for a fairly large number of observed facts and is found to be of universal application is generally called a theory.

It should be noted, however, that a hypothesis is not a necessary part of scientific work. Experimental facts can stand by themselves whether there is a hypothesis or not to account for them. It is also necessary to remember that a scientist need not do any experimental work for the formulation of a scientific theory. A theory can be formulated on the basis of experimental work done previously by a number of other workers and the data which are already available.

The Theory of Relativity is a remarkable illustration of this kind of formulation. On the basis of the results obtained by a number of scientists in different fields of work and the data and ideas provided by them, Einstein worked out a theory which is perhaps the most remarkable and of far-reaching consequences in the history of Science. With only a pencil and paper and his theoretical knowledge of Science he fused together a tremendous amount of experimental knowledge scattered in different fields of Science into a theory which has not only revolutionised scientific thought but produced a great impact on the life of humanity as a whole. It is true that his theory was also tested by some crucial experiments but this experimental corroboration came much later.

What has been said above will show that it is not necessary to be an experimental scientist to be able to throw light on scientific problems and affect scientific thought. A broad, general but accurate grasp of scientific knowledge combined with a clear comprehension of Occult doctrines and capacity to see deeper into things can enable a student of Occultism not only to understand both more thoroughly but also to throw light on both and bring them into a more meaningful relationship. A student of Occultism has the great advantage that he can rely implicitly on the Occult doctrines, at least in their broad outline, for reasons which will be discussed later on. He therefore stands on *terra firma* among the shifting sands of scientific thought and can thus provide a permanent light-house for those who are seeking some reliable standard of reference by which to steer their course in life among the surging and confusing billows of modern thought. This is really a duty which devolves upon those who have the necessary intuitive perception to be able to see the significance and value of this precious knowledge which has been vouchsafed us in the outer world.

The scientist tries to investigate the ocean of unknown facts and on the basis of the very meagre data which he has been able to collect, draws his conclusions with regard to everything, right from the mechanism of sensuous perception to the origin and nature of the universe. The paucity of this data can be judged from the fact that he has not only to deal with an infinite number of facts on the physical plane with his limited resources, he has also, out of sheer prejudice, placed out of bounds for himself the whole realm of mind and consciousness and all the subtler worlds which are hidden within the shell of the physical universe. Facts of existence do not cease to exist when people ignore them, even though these people may be the scientists of the modern age. So when a scientist draws his conclusions the infinite realm of both physical and super-physical facts always remains unexplored or partially explored in the unlimited background of the unknown. It is this fact which makes it inevitable that he can never arrive at the final truth of existence and cannot be quite certain even about the partial and limited truths which he has discovered. The discovery of new and unexpected facts, not even suspected by him, may invalidate his conclusion at any time as has been seen again and again in different fields of scientific investigation.

As long as the scientist confines himself to facts experimentally discovered by him he is on very firm ground and no one can dispute his claims or blame him for drawing his limited conclusions based on these partial facts. It is only when he leaves his legitimate domain and on the basis of an extremely insufficient physical data and pre-conceived notions pronounces unwarranted opinions on the ultimate problems of human life and the universe, that he becomes liable to gross errors and the charge of prejudicing the human mind and thus interfering with the natural course of human evolution.

It will be seen from what has been pointed out above about the scientific method that it is inherently incapable of discovering the fundamental truths of existence in their 'real' form. This is so not only on account of the limited nature of the inductive method adopted by Science in its methods of investigation but also because scientists have arbitrarily and deliberately placed before themselves an extremely limited objective by confining themselves only to physical phenomena and physical methods in their investigations. They have decided to ignore the vast and more important fields of mind and consciousness which undoubtedly form the background of the physical phenomena and are an indispensable factor in the investigation of these phenomena. The 'observer' comes into the picture in every field of scientific investigation and without him there will be no scientific work, and yet he is ignored in the most irrational manner and scientific theories are formulated and based upon purely physical facts as if mind and consciousness had no factual and independent existence and did not count.

Scientists, especially those whose orthodoxy is less rigid, will probably deny the above charge. But the attitude which scientists as a whole adopt towards the doctrines of Occultism, and the fact that in their theories concerning the nature and origin of the universe, mind and consciousness do not enter at all is an indirect proof of the above charge.

Mind and consciousness are subjective realities which cannot be observed and investigated by objective methods and therefore Science brings forward the excuse that its methods are objective and it therefore confines itself to purely objective phenomena which can be investigated by objective methods. But what about the subtler worlds which are hidden within the physical world? These worlds are objective in nature, can be observed and their phenomena investigated when the subtler senses which are lying dormant in every human being have been activated. All seers, mystics, occultists and great religious teachers of the world have testified in unequivocal terms to the existence of these worlds. Societies for psychical research have investigated many abnormal psychical phenomena and come to the conclusion that there is at least strong probability if not certainty of such worlds existing. In the face of this overwhelming and reliable evidence with regard to the existence of these worlds what has Science done to investigate whether such worlds do actually exist, and if they exist, what is their nature? Absolutely nothing. To say that these superphysical worlds do not exist because they cannot be observed with physical apparatus is obviously irrational. It shows that modern scientists have assumed and decided that such worlds cannot and do not exist and therefore lay down impossible conditions for their investigation as an excuse for doing nothing.

The fact probably is that most scientists fear, in their heart of hearts, that the whole structure of scientific thought based arbitrarily upon a purely materialistic philosophy will come tumbling down the day the existence of these superphysical worlds is established beyond doubt, and mind and consciousness are definitely proved to be independent realities in existence. It is this fear which lurks behind the imposing facade of scientific attitude that is really responsible for the fact that they are not prepared even to look at all these superphysical matters.

There is something fundamentally unscientific in an approach which sets out to discover the ultimate truth and then stops short at a particular stage in the process of discovery, declaring: 'Thus far and no further'. The seeker after Truth must dare to face all facts and not shy at those which are at variance with his preconceived notions or interfere with his favourite pursuits. He must be prepared to give up everything, face all kinds of difficulties, sacrifice not only all personal comforts and pleasures but also his cherished ideas and favourite pursuits. Those who found the ultimate Truth of existence fulfilled all these conditions in their search for Truth and that is why they succeeded, and those who want to find the same Truth again for themselves must follow in their footsteps. If any one thinks that it should be easy to undertake this task and find the ultimate Truth—if there is any such Truth in existence—let him read two books in Occult literature—*Light on the Path* and *The Yoga-Sūtras* of Patāñjali—and then ask himself how many people among the modern seekers after Truth will be willing to go through this kind of self-discipline or are capable of making such a prolonged and determined effort.

Although we are concerned in this book particularly with the relation of Occultism and Science it is necessary also to review here very briefly the methods of enquiry and discovery upon which Philosophy and Religion are based. For, these three branches of knowledge—Religion, Philosophy and Science—and their methods of enquiry and investigation are intimately related to one another and can be considered as complementary in the highest conception of this search for Truth. Failure to recognize the above fact is responsible for the mutual suspicion, exclusiveness and working in water-tight compartments which characterize modern Philosophy, Religion and Science. This exclusiveness, which is based on sheer prejudice and narrowness of outlook, has unnecessarily deprived philosophers, scientists and leaders of religious thought of the great advantage of mutual exchange of thought and collaboration and made their respective methods less effective than they need have been. If Truth is one and indivisible then the methods of enquiry and discovery of that Truth should also be one and indivisible and it should be impossible to arrive at the ultimate Truth by following rigidly one particular method. Let us now consider the discipline of philosophy.

As induction is the instrument of discovery in the case of Science, so deduction—proceeding from the general to the particular—is the method of enquiry and discovery in the case of philosophy. Academic philosophy makes certain basic assumptions with regard to the fundamental facts of existence and from these basic assumptions elaborates, with the help of reason and common experience, a well-knit and plausible system of thought which can explain, as far as possible, all important facts of human life and the universe. Considering the complicated nature of these facts, the uncertainty and the arbitrary nature of the basic assumption and absence of any touch with the inner realities of existence such a philosophy could never hope to come anywhere near the Truth as its conclusions could be merely wild guesses. But most of the great philosophers were either men of some mystic or occult experience or borrowed their basic ideas from the common doctrines of the great religions. That is how their philosophies, in their broad conclusions, generally reflect at least to some extent the fundamental truths of existence.

The weakness of the deductive method which is based merely on reason and common sense is thus the uncertainty and arbitrariness of the fundamental postulates. The conclusions arrived at by reasoning will vary according to these assumptions and one can manage to arrive at any predetermined conclusions by working backwards and making suitable basic assumptions. You can start with the idea that there is no God and have a purely materialistic philosophy or assume that there is a God and have an equally reasonable theological philosophy. Academic philosophy is thus to a great extent a matter of guesswork and its fundamental defect is that its motivation is the production of a very clear and plausible system of thought and not the discovery of the real truths underlying phenomenal appearances.

In considering the deductive method of discovery we have to take into account the intuitive faculty of man. As man is a microcosm containing within himself, in a potential form, all realities in existence, all truths are hidden in a subtle form within his consciousness and it is, therefore, possible to perceive these truths intuitively at least in a partial form. Whenever and wherever the human mind becomes attuned, owing to any cause, with the Universal Mind a flash of the underlying truth can illuminate the whole field of enquiry. The truth is obtained in a flash from within as an intuitive perception or wells up slowly and mysteriously as integrated knowledge and is then elaborated and formulated through the intellectual apparatus for presentation to those who have not had the benefit of the direct vision. In this transition it loses much of its clarity, directness and vivid quality and appears as a laboured and reasoned proposition which may or may not convince or be acceptable to those to whom it is presented, although this question cannot possibly arise in the case of the original recipient.

There is yet another method of gaining knowledge concerning the deeper realities of existence. This is through initiation when knowledge is imparted from one who knows, and has the necessary power and authority to do so, to a qualified disciple by temporarily raising the receptivity of the latter by occult methods. Owing to this artificial sensitivity induced in this manner the consciousness of the recipient is temporarily raised to a higher level and is able to perceive the particular truth as a direct experience in consciousness.

This kind of sensitivity does not, however, last and is not under one's control, as it has been induced temporarily by an external agency by the descent of particular kinds of forces in the vehicles by esoteric methods. But it generally sets in motion processes which culminate in the capacity to gain these realizations voluntarily and by one's own efforts. This is really the realm of Yoga and we need not enter into this question further in the present context.

So we see that the value and validity of a particular philosophy will vary with the nature of its origin, the capacity of the propounder and the manner of its formulation. Its nature may vary from a wild guess based on speculation to a fairly satisfactory presentation of the inner realities based on their unveiled spiritual perception. All these philosophies are there claiming our attention and our allegience, if possible, with nothing external to indicate their value or validity, and the aspirant for the Truth has to exercise his discrimination in judging their value and authenticity. His capacity to exercise his judgment correctly and distinguish between the true and the false properly provides the first test of his ability to tread that inner path of unfoldment of consciousness which leads to the discovery of the ultimate Truth of existence.

It will also be seen that the value of the deductive method depends to a very great extent upon the correctness of the basic assumptions and how far they reflect the fundamental realities of existence. If these assumptions are correct, at least to a reasonable extent, the conclusions drawn from them by a process of reasoning will also be correct and proper, and provide not only a reliable philosophy of life but also an efficient guide to conduct, both to the individual and society. If they are arbitrary and incorrect, based upon prejudices, personal whims and reactions to surrounding conditions, and not upon any natural laws, outer or inner, then they will naturally produce a philosophy of life which not only does not truly represent the inner realities of existence but leads to more and more difficulties, conflicts, frustrations and chaos.

If, for example, we base our philosophy of life on the well-known Occult doctrine that all living creatures share a common life and are one in essential nature, then there should be no real conflict of interests between different individuals, communities or nations and mutual sympathy, help and co-operation can be the only right basis of our relations in everything. A complete guide to human conduct and a just and humane social order based upon mutual help, consideration and co-operation can be deduced from such a fundamental principle and such an individual life and social order will lead to harmonious relations, peace, stability, contentment and a rapid development of the best side of human nature because it would be in harmony with the inner realities which form the basis of the universe. On the other hand, if we base our philosophy of life on the materialistic doctrine that human beings are the chance products of the interaction of blind and insentient forces of Nature, that they have no reality beyond the physical life, that competition and pursuit of self-interest is the natural law of survival, then inevitably such a philosophy must lead to conflict of interests, rivalry, exploitation, constant struggle and instability, with their natural consequences, fear, unhappiness, anxiety and general chaos.

So the question whether the spiritual or the materialistic philosophy of life is the true and correct one is not merely academic, a problem of philosophy. It is a question upon the solution of which the progress, happiness and the very survival of humanity depends. It is not necessary to prove what has been said above. We have only to open our eyes and see what is happening in the world as a result of the wrong premises upon which our present philosophy of life and social order is based. Even those who profess a spiritual outlook are guided in practice by the materialistic philosophy. A philosophy of life can be judged easily by its fruits.

The approach of religion to the deeper problems of life is quite different and is based upon revelation. The orthodox followers of a religion generally take for granted that the Teacher or Prophet who promulgated the religion knew the Truth and therefore the conception of God or man and the code of conduct which he has placed before them is the correct one and there is nothing further to be known and done in the matter. All that they have got to do is to follow the teachings of the Teacher as embodied in the sacred books of the religion, to the best of their ability, and hope for the best as far as their future life is concerned. There is no place in such a religion for experiment and induction, as in the case of Science, or for assumption and deduction as in the case of Philosophy.

This is not the place to discuss the tenability or value of this approach but a few relevant facts may be pointed out very briefly to enable the student to understand properly the essential nature of this approach and its real role in the discovery of Truth. These may be stated as follows:

(1) The great religions of the world which have been promulgated in the past have been given out with the main purpose of providing a guide to conduct for a particular people living in a particular set of circumstances and to enable them to learn the elementary lessons of spiritual life.

Their main purpose was not to provide knowledge concerning the inner realities of life and training to enable aspirants to gain direct knowledge of these realities. This task can be taken up only by a few souls who have gone through all the necessary experiences of life and thus developed a certain degree of spiritual maturity. If, however, there existed the possibility of a fairly large number of individuals being present who needed this higher knowledge or training these were provided separately by the Teacher but these generally remained esoteric and not mixed with the teaching for the common man.

This does not mean that real aspirants and candidates for the Path of inner unfoldment were left at any time without knowledge or training if they needed these for their further progress. There have always existed in the world secret schools of mysticism and occultism or individual teachers of high spiritual development who could impart such teaching and guidance to the few who needed such help and were properly qualified to receive it, and those who need such help and guidance always manage to find their way to them.

(2) The possibility of revelation of truth is a fact of Nature and history. Humanity has not to climb from the bottom to the heights of spiritual knowledge by its own efforts from the very beginning. From time to time this knowledge is dropped from above to enable humanity to climb to the heights of spiritual knowledge by its own efforts. First individuals, then humanity as a whole learn to undertake this strenuous and difficult task and find the Truth which is hidden within their heart by direct perception gained by their efforts. Revelation is a part of this preliminary training.

Occultism has always taught that spiritual knowledge needed at a particular stage of development can and does come from above. In fact, this is necessary in the earliest stages of the evolution of humanity when there are no individuals sufficiently evolved in it to bring down the 'fire' of heaven from above. Spiritual teachers from other humanities which are more evolved guide the evolution of our humanity from behind the scenes and appear among different peoples at different times to impart the teaching which is needed at the time. Or, a direct descent of Divine Life takes place at critical stages of evolution to impart the initial impetus needed for this purpose. It is this fact which is embodied in the well-known doctrine of *Avatāras*, *Avātara* in Sanskrit meaning 'descent from above'.

It is not only in matters of religion that this kind of help comes from above. In other fields also there is constant flow of ideas, knowledge and other evolutionary forces through individuals who are able to transmit such knowledge in this manner. The manner in which scientific knowledge and technology have developed during the present century shows unmistakably the hidden hand which stimulates and guides this phenomenal development. But our modern 'intellectuals' will resent the very idea of being made instruments of higher powers in this manner and it is useless to expect them to believe in such 'superstition'.

(3) It is also necessary to remember that the original teachings of the great religious teachers, partial as they are, are also liable to be changed and debased as they pass through the minds of their followers century after century. This is partly due to natural deterioration which sets in with passage of time but is chiefly caused by misinterpretation by those who have completely lost touch with the realities of the inner life and can see everything only with the eyes of the flesh. And when the desire to exploit these teachings for

their narrow selfish purposes is added to all this we are left only with dead creeds which most people follow mechanically to fulfil their religious duty. So, in the case of religions also, as in the case of philosophies, it is necessary for the serious student and aspirant for truth to be on his guard and exercise his discrimination constantly to separate the true from the false knowledge which covers the kernel of truth in every religion.

It has been pointed out already that the ultimate Truth underlying the universe of different degrees of subtlety must be whole and indivisible and therefore the methods of discovering it must also, by their very nature, be closely inter-related and indivisible. The artificial division of religion, philosophy and science which has been created and maintained by their respective votaries thus does violence to this fundamental principle and by creating unnecessary barriers between these three basic branches of knowledge renders their fundamental task of discovery of Truth extremely difficult, if not impossible.

Is there any branch of knowledge and method of discovering truth which is not hampered by these divisions and inhibitions in the pursuit of Truth? Yes. Occultism. And it is because Occultism has kept itself free from these restrictions that it has succeeded not only in discovering the Ultimate Truth but in keeping open that road which leads to its discovery. For, it is when intellectual knowledge loses contact with the realities which it is meant to reflect that it becomes corrupt and distorted out of shape. The only way of preserving such knowledge pure and fresh is to keep the channel between the spiritual knowledge and intellectual knowledge open, so that there are always present in the world a number of enlightened individuals who are permanently established in Truth and their ranks are being constantly swelled by fresh entrants who have trodden the Occult Path and succeeded in gaining direct knowledge of the inner realities.

It is these conditions which have made Occultism the only reliable and permanent source of true knowledge with regard to the inner realities and the Occult Path the only means of finding and knowing these realities by direct experience. Scientists ought to be able to understand this principle, for the reason why experimental scientific knowledge, extremely limited as it is, is kept pure and fresh and can be relied upon under all conditions, lies in the fact that there are always present experts who have direct knowledge of the facts and there are others who try and succeed in verifying these facts.

It is true that all the knowledge of Occultism, unlike scientific knowledge, is not available to the general public; that is why it is called Occultism, *Gupta Vidya*, or ' the Secret Knowledge '. But this is so, firstly, because most of it, in its real form, is beyond the comprehension of even the highest intellectuals who are still confined within the prison-house of the intellect. And secondly, because it confers tremendous powers on those who possess it and are bound to misuse these powers if they are not morally and spiritually qualified to wield these powers. It is not necessary to prove this point. The unrestricted and irresponsible misuse of powers conferred by scientific knowledge of merely physical forces shows what risks will be involved if these people are able to get hold of psychic and spiritual powers related to Nature's finer and tremendously more powerful forces. It is for this reason that the Adepts of Occultism who possess and are the custodians of this Supreme Knowledge have raised only a corner of the veil which hides the unlimited and inconceivable vistas of Occult knowledge, so that those who are ready and mature enough may get a glimpse of the glories which lie beyond and make the necessary efforts to tread that difficult Path which will ultimately qualify them to penetrate through the veil and pass into the real worlds which lie beyond.

It is because on the path of Occultism it is possible to discover the ultimate Truth in which all other truths are contained in essence that the factual knowledge embodied in the Occult doctrine, revealed or unrevealed, is final, not subject to change or correction and therefore utterly reliable. No new facts can emerge from the unknown to invalidate it in any respect because there is nothing else left to be known. It is not that new facts cannot be discovered, but these new facts are always in the framework of the known facts and are an integral part of the ultimate Truth.

It is because the fundamental assumptions upon which the Occult philosophy is based are correct and derived from direct experience that this philosophy, even in its intellectual framework, is substantially correct and therefore reliable. There might be minor variations from the Truth as it really exists but these can be due only to the faults of the intellectual apparatus through which they have passed in interpretation and presentation. There can be no major or fundamental error caused by erroneous assumptions. Generally, when there are apparent contradictions they will be found on closer analysis to be due to looking at the same truth from different points of view.

Again, taking the religious aspect of Occultism, we see that the Path of Occultism leads to true, direct and highest religious experience and therefore not only frees the individual from the illusions and miseries of life permanently but also gives him that Peace which passeth understanding. Religious life based on faith can crumble at the touch of adversity or temptation but that grounded in true religious experience cannot. This Truth which is sought to be known in the highest religious experience, and is generally referred to as God, is the same Truth which the philosopher seeks through penetrating discrimination or *Viveka* and the scientist seeks through experiment. The philosopher, scientist and the mystic seek the same Truth although they call it by different names and adopt different but complementary methods for its realization.

Having pointed out the identity of their apparently different goals let us now try to show how the method of Occultism combines the apparently different methods of achieving that goal into one integrated whole and thus establishes also the essential identity of method. This principle of the essential identity of method follows directly from the most fundamental Occult doctrine according to which there is only one Ultimate Reality in existence. For, it can be seen easily that if there is only One Truth which Religion, Philosophy and Science seek under different names and expound in different ways, then the methods of discovering that Truth in the three disciplines must also be related to one another and should be merely different aspects or expressions of one integral method of approach.

We have seen above how induction, deduction and revelation are the predominant characteristics of the methods adopted by Science, Philosophy and Religion respectively in the discovery of truth. Let us see what kind of role these three methods of gaining knowledge play in Occultism. Let us take induction first. Induction, as pointed out already, involves the collection of experimental data which serves as a basis for drawing conclusions and making further enquiry and experiments. What part does induction play in Occultism? In order to answer this question let us consider what is the fundamental nature of the facts upon which the Occult method of enquiry and investigation into the realities of life is based and from which it starts. These facts are all connected with the nature of human life and human experience in their essential nature and were set forth in a very simple manner by the Buddha in 'the truth of sorrow' upon which his eightfold path is based. Every spiritual teacher starts from this point and proclaims in no uncertain terms that misery, illusion and limitations are inherent in the very nature of this phenomenal world and the only means of getting out of these conditions is to find the real world which is hidden within this phenomenal world. This fact has been stated very clearly and tersely in apphorism II-15 of the *Yoga-Sūtras* which shows why it is necessary to tread the path of Yoga outlined in that book. The aphorism sums up all human experience and draws an obvious conclusion from the result.

It will be seen therefore that those who took up this enquiry into the nature of Reality and the ultimate Truth of existence did not start on the basis of a few facts related to physical forces and laws of nature and collected through physical apparatus. They surveyed the whole field of human thought and experience in one wide, all-embracing sweep and detected a few universal facts which could not be disputed, and were of the utmost significance and importance to man. They saw that everything in this phenomenal world was temporary and subject to change and therefore no stable foundation of security, peace, and happiness could be built on this incessant flux of phenomena. They found that everything in this phenomenal world was permeated with illusions of different kinds and degrees and the senses upon which all our physical life and experience depends are deceptive and the knowledge obtained through them is therefore worthless. Science working on the same physical data for a century has also at last discovered the deceptive nature of the data collected through the sense-organs or physical apparatus, but in its desperation to avoid facing facts of existence squarely it has now taken refuge in a reality of mathematical abstractions.

But those who blazed the trail of investigations into the realities of life arrived at this important conclusion only on the basis of ordinary human experience with the help of their penetrating discrimination. And they made up their mind to find the reality behind these ever-changing phenomena irrespective of whatever sacrifices they had to make, whatever methods they had to adopt, and however long was the time it took to arrive at the ultimate Truth. That Truth and nothing but that Truth was their sole objective.

Compare with this the attitude of those who profess to find the truth underlying the universe by the scientific method. They now realize the ever-changing nature of the phenomenal world and the illusions created by the senses and yet they ignore this fundamental fact of existence and continue their work as if this was the only real world in existence. They profess that they want to find the truth underlying the universe but they will use only physical methods which are in accord with their preconceived notions about the universe. They are trying to discover the ultimate basis of this physical world and yet they insist on ignoring mind and consciousness which are not only the most important and significant realities in human life but the most important factor in the observation of even physical phenomena. Does this kind of attitude show that they are seekers after Truth and want to find the Truth and nothing but the Truth, and is it possible for those who adopt this extremely limited and prejudiced approach to the problem, to find the unlimited Truth of existence?

The second essential aspect of the inductive method is experimentation. Having seen that mind and consciousness play the most important role even in the phenomenal world, the original seekers after Truth naturally concluded that the Great Secret could not be found in the purely physical world and the Reality hidden beneath it could not be discovered by purely phenomenal methods. You cannot destroy one illusion by another illusion as you cannot destroy darkness by darkness but only by light. So they decided not only to transcend the physical world perceived through the deceptive sense-organs but also the subtler mental worlds perceived through the mind, for the mind was also seen as a creator of illusion.

Having arrived at this conclusion and adopted the course of action indicated by it, they experimented with their mind in the most determined and intensive manner and succeeded in uncovering layer after layer of mind and the mental worlds created by it until they discovered the world of Consciousness or Reality hidden behind the subtlest level of the mind. The whole Science of Yoga was laboriously and patiently built up in this manner with the help of the Divine Life which always guides such efforts from above. This Science enabled the pioneers in this field not only to find the Truth for themselves but also to provide a systematic and reliable method of discovery of this Truth for others who are prepared to embark on this voyage of Self-discovery and undergo the self-discipline needed for this purpose.

The Occult method of discovering truth which has been outlined above will be seen to involve the inductive method in its highest form. It not only involves experimentation to find facts but experimentation undertaken with no reservations and prejudices about the nature of the objective and the means of gaining that objective—which cannot be said about the modern scientific method of discovering the reality behind the phenomenal world.

What about the role of deduction in the discovery of truth by the Occult method? The first remarkable achievement of this deductive method was the chain of conclusions at which the seekers of Truth arrived as a result of the observation of universal misery, suffering and limitations inherent in human life. The conclusion that the phenomenal world cognized by the senses is illusory and therefore not to be taken seriously as a basis of our happiness and object of detailed investigation is the first link in this chain. That mind and consciousness which form the background of the phenomenal world and must therefore contain the secret of the universe was the second important conclusion. That only mental methods could unravel the mysteries of mind and consciousness was the third. With these fundamental conclusions as their premises they started experimenting with their life and mind and by persistence and courage discovered, one by one, the inner realities of life and finally the Ultimate Reality.

But deduction plays another important role in Occultism. Having discovered the deepest and most fundamental truths by direct perception it is possible to deduce other minor truths and laws from these truths by a process of reasoning and intuition. This universe is extremely complex both in its noumenal and phenomenal aspects and occult knowledge of the fundamental realities and principles cannot provide us an answer to every problem and question which we have to deal with in our life. Most of these problems can however be solved on the basis of the fundamental principles by a process of reasoning guided by intuition. The fact that the manifested universe is the result of the repeated projection, expression and reflection of the One and the same Reality at different levels of manifestation is also of great help in this task on account of the analogous relations being established between the lower and higher expressions of the same principle. The mathematical basis of the universe is yet another fact which can be utilized very èffectively in this work. Many examples of this kind of deduction will be found in *Man, God and the Universe* and many more will be found in the present work in discussing different problems of modern Science.

It is not implied that we have to depend entirely on deduction with regard to the discovery of these minor truths. Direct and certain knowledge with regard to any matter can be obtained at any time by those who have developed the necessary capacity for this purpose. But it is not necessary to resort to these more difficult and esoteric methods in everything if we can get the same result by more easily available and universally applicable methods of deduction. The intellect cannot lead to wrong conclusions which contradict facts of direct experience provided the basic premises are true, the mind is pure and unprejudiced and the reasoning is correct. After all, it is the expression of the *Chit* aspect of the Self and the instrument of soul at the lower levels of manifestation.

As an illustration of application of the deductive method in Occultism we may take the field of ethics and social organization. From the basic principle of the One Life which follows from the doctrine of the One Reality it is possible to deduce in outline not only a reliable and workable science of ethics but also of social organization and international relationships. This individual morality and sociology will naturally be based upon love, compassion, justice and brotherhood in the true sense because it is these qualities which truly reflect our real mutual relations and keep us in tune and harmony with the fundamental Reality underlying the universe. And it will try to eliminate, as completely as possible, from our mutual relations in every sphere, hatred and all its offshoots such as cruelty, exploitation and violence. Hatred being the negation of love and therefore against the essential reality of our mutual relations is bound to create more and more serious problems, increasing conflict, all ending in universal chaos and disaster, unless its devastating consequences are counteracted in time, at least by justice and undoing of wrongs, if not by positive policies based on love and brotherhood.

A critical and impersonal examination of the conditions prevailing in the world can easily provide an explanation of the chaotic and dangerous situation which has been created by a sudden and enormous accession of physical power in our hands with practically no further unfoldment of our wisdom. It will be seen to be the inevitable result of the wrong ideals and policies being pursued in the sphere of national life as well as in international relationships. What is wrong with these schemes which have been drawn up so elaborately and enforced so ruthlessly by our modern planners? They are masters of logic and reasoning and have achieved spectacular success in different fields of scientific technology as a result of this faculty. Their premises are all wrong. They assume that there are no laws operating in the realm of mind and consciousness while in reality every aspect of the manifested world is governed by natural laws as in the case of the physical world. A corollary of the above assumption is that man is free to do what he likes as far as his mental and social life is concerned and has to take into account only the laws of the physical world. But the fact is that man is strictly responsible for everything he does both as an individual or as a community, and he has to bear with mathematical exactitude the consequences of what he does, sooner or later. Another basic assumption in the philosophy of scientific materialism is that there are no other worlds except the physical and the total man is merely a product of physical forces working in his body and outside. But the real fact is that the physical body of man is merely the outermost vesture of the real man existing on all the subtler planes of being, and subtler forces of all these planes are constantly influencing and moulding his life all the time. He not only exists on these subtler planes but also shares a common life and a common consciousness with all other human beings on the spiritual planes. The brotherhood of man is thus not a mere ideal artificially created by idealists or politicians but a fact in nature which must be taken into account in our life and mutual relationships.

So, it is these errors in the basic assumptions of the philosophy of scientific materialism which are responsible for the conflicts, violence, fear, chaos and instability which are the normal features of modern civilization. Even those who profess to be against this philosophy follow it in practice and thus share the responsibility and blame for these conditions. It is not profession but practice of what is right and in accordance with natural laws which brings about right conditions as a natural sequence.

What role does revelation play in the search for Reality by the Occult method? It has been pointed out already that revelation is essentially the descent of knowledge from a higher source into a receptacle at a lower level. There are two ways of gaining knowledge concerning the realities of the higher and subtler planes which are beyond our normal powers of perception. One is the method of Yoga in which the Yogi himself brings about the necessary changes in his mind and vehicles to enable his consciousness to expand and establish contact with the higher planes of being in *Samādhi*. The other is the method of revelation in which knowledge is imparted by a highly advanced spiritual teacher to a disciple which he cannot gain by his own unaided efforts. This is possible because there are certain centres and organs within the vehicles of a human being whose stimulation enables consciousness to expand temporarily and become aware of subtler realities, or to receive knowledge and messages clothed in language through the ordinary process of hearing or seeing. The temporary stimulation of the organs or centres depends upon the manipulation of certain forces in the vehicles of the recipient, and can be brought about only by a highly advanced teacher.

Revelations of this nature are a normal feature of the path of discipleship which leads to Liberation. Orthodox people generally associate revelation exclusively with 'prophets' or religious teachers who have founded the religions which have appeared in the world from time to time. But that is taking a very narrow and restricted view of revelation. In fact, if we consider 'Initiation' as a form of revelation it has to be considered as an integral part of the process of spiritual unfoldment. Initiation, as is well-known, involves the intervention of a higher agency in the process of descent of spiritual knowledge and power for a limited purpose.

According to Occultism revelations of spiritual truths play a very important part in the early stages of evolution of every humanity when the people themselves are incapable of finding these truths by their own efforts but need them for their further progress. It is for this reason that some of the highest and purest forms of teachings concerning the truths of the inner life are found among the scriptures of the earliest races which were not evolved enough to bring down this knowledge by their own efforts. These teachings were given either through a religious movement or through a line of teachers and disciples who pass on this knowledge from one generation to another and systematically train disciples to establish ideals, schools of self-discipline and religious centres in the community on a stable basis. Many schools of Yoga and religious sects in Hinduism which have lost their original dynamism and spiritual knowledge trace their descent to such schools founded generally by a well-known historical or legendary Teacher.

What has been said above will show that the three methods required in the expansion of knowledge—induction, deduction and revelation—can not only be employed together but that such an integrated approach to the problem is necessary for an all-round, harmonious growth of true knowledge. The exclusive use of the scientific method without the correcting and balancing influence of religion and philosophy leads more and more to the abnormal growth of purely empirical knowledge which progressively limits man's vision and power of independent thinking and makes society depend more and more on technocrats. A modern technocrat is really a man who has put on mental blinkers and is incapable of seeing correctly or judging properly the larger and more important issues of human life in the correct perspective. A society guided and controlled by technocrats is like a crowd whose members are putting on blinkers and colliding constantly with one another in their movements. Modern society, with more and more stress and the necessity of specialization, is heading towards a social order guided by technocrats in every field and a time may come when technocrats control all spheres of our life, and democracy is replaced by technocracy as envisaged by H. G. Wells.

Leaving aside the developments which may take place in the future, even the present prospect is not very pleasing either. Military technocrats decide whether we shall live or die. Political technocrats called politicians decide whether we shall be free or slaves, whether we shall fight or live in peace. Biologists decide whether we shall consider ourselves as animals or as Divine beings. Physicists and astronomers decide whether there is a God or no God. Modern physiologists and psychologists decide whether we have really an independent mind or see only the results of the biological reactions in our physical body as mental phenomena. If this trend continues undetected and unchecked and the individual ceases to think for himself gradually, and hands over the guidance of his life to technocrats, we shall be reduced to a slavery of the worst kind, all the more insidious because it is self-imposed and has the veneer of high civilization.

What has been said above about the isolated growth of scientific knowledge holds equally true of knowledge in the realms of philosophy and religion. Philosophy divorced from facts of human experience and the real issues of human life tends to become more and more speculative and degenerates more and more into mental hair-splitting and playing with ideas as a pastime or profession. Religion divorced from experimentation, experience and direct knowledge as well as from a wide philosophic outlook tends to become a meaningless repetition of the same old dogmas and pious religious sentiments, and a mechanical performance of routine religious duties.

It is for this reason that Occultism advocates an integrated approach in which science, philosophy and religion play their respective roles and make their respective contributions in the growth of true knowledge and the proper and harmonious unfoldment of human mind and consciousness, both in the individual and society. And it is for this reason that the Occult path combines the special techniques of science, philosophy and religion in the discovery of Truth. The Ultimate Reality is an integrated state and can be realized only by adopting an integrated approach and employing an integrated technique of discovering Truth.

This book is not meant to deal with the relation of Science and Occultism in an exhaustive manner or even in detail. Our knowledge of genuine Occultism is incomplete and fragmentary; firstly, because most of it is beyond the comprehension of the human intellect, and secondly, even what is comprehensible has not all been made public in the conditions prevailing in the world at present. Whatever has been given out is of a general nature and cannot, as it is, throw much light on the precise and specific problems of modern Science. It has to be properly understood and assimilated before it can be used in this manner. But even a general acquaintance with the Occult Doctrine with an open mind and intuitive insight is enough to show the student that here we have a fund of knowledge of unfathomable depths which holds within its deeper recesses the innermost secrets of the universe and can throw some light on many difficult problems of modern Science.

If we study carefully and with discrimination these ancient doctrines of Occultism which are found scattered in genuine Occult literature and also acquire a general knowledge of Science we can bring the two together in some kind of meaningful relationship. This will help us to understand more easily the doctrines of Occultism in the light of modern scientific ideas and also to throw some light on the theories of Science and enable us to see their deeper significance. And it is possible that some men of Science who are not very orthodox and rigid in their outlook may become inclined to examine these ideas for the solutions of many problems which they are facing at present. More than this we should not expect from them at present.

A lot of research work by students of Occultism is needed for opening up lines of communication between Occultism and Science. If those who have adequate knowledge of Science take up this work in the true spirit of enquiry and to promote the cause of the Occult philosophy they will find that this is really a very fruitful field of research which will help us not only to understand the doctrines of Occultism much better but also to create the necessary atmosphere in the world for their acceptance.

Theosophy is generally considered as an ancient philosophy which may be interesting from the academic point of view but has no particular relevance to the pressing problems of the modern world. The only way in which this impression can be gradually removed is to enter the current of modern thought through research work of the kind mentioned above and to show that these eternal truths have not only relevance for the problems of the modern world but can provide generally very satisfactory solutions of these problems. This book has tried to show how a beginning can be made in this direction.

If it can encourage others to enter this field and take up this kind of work in right earnest it will have served its purpose.

The student who goes through this book may feel that there is some repetition and over-lapping of ideas in dealing with different topics. It may be pointed out that this is to a certain extent unavoidable in a book which tries to throw some light on different aspects of one and the same Reality of unfathomable depths and countless aspects from different points of view. This could be avoided only by means of frequent and elaborate cross references which cause dislocation of the train of thought and make the study of a subject unnecessarily tedious. The author has tried to make each chapter complete in itself as far as possible, to give a clear and coherent idea of the subject dealt with. The repetition of ideas which this has necessitated will perhaps help in further clarifying those ideas and their being assimilated by the mind of the student.

The relation of Science and Occultism on which this book is meant to throw some light is indicated to some extent by the following well-known quotation which shows clearly the necessity of our investigating in the Theosophical Society the common ground between these two important branches of knowledge and trying to bring them into closer and more meaningful relationship:

"The doctrine we promulgate being the only true one, must, supported by such evidence as we are prepared to give, become ultimately triumphant as every other truth. Yet it is absolutely necessary to inculcate it gradually, enforcing its theories, unimpeachable facts for those who know, with direct references deduced from and corroborated by exact Science."[1]

[1] *Letters from the Masters of the Wisdom* first series, (1973) p.2, The Theosophical Publishing House, Adyar, Madras.

CONTENTS

CHAPTER I

RELATIVITY—TOTAL AND PARTIAL—I

Scientific Theories and Occult Doctrines

Einstein's Theory of Relativity is considered to be the most remarkable contribution to scientific thought in this century. Although it was based on the researches of a number of topmost scientists and mathematicians who had made very important and fundamental contributions in their respective fields of work from time to time, Einstein's brilliant work consisted in fusing together all these scattered and unrelated ideas into a well-knit theory which has given a new trend to scientific thought and stimulated research of a very fundamental nature. Although the theory was hailed as a magnificent contribution to scientific thought from the very beginning, very few people understood it in the early stages owing to its highly abstract and mathematical nature. And so it remained for a long time a piece of scientific work much admired but little understood. It was only with the spectacular discovery of atomic energy which was predicted by the theory, that interest in the theory was suddenly stimulated, its various conclusions began to be tested and accepted widely. As always happens in such cases the sudden tide of popular acclamation swept everything aside and the theory has come to be considered as the bible of orthodox scientific thought.

It may be worth while digressing here for a while to point out that although the work done in the field of Science, especially, in its technological developments, is truly wonderful and one can have nothing but profound admiration for the large number of scientific workers who have contributed to these phenomenal developments, still, it is necessary to keep in mind the Occult doctrine concerning human evolution. According to this doctrine the lower worlds of illusion are mere shadow worlds which merely reflect the Divine Plan in their evolution. As the Divine Plan unfolds, what is in that Plan begins to find expression in the lower worlds automatically through the most appropriate instruments and conditions which are available. The conditions are prepared as thoroughly as possible beforehand but the process cannot be interfered with by any defect in these conditions, just as the projection of a cinematographic picture cannot be interfered with by any defect in the screen on which it is projected.

Under the pressure of this descent of forces, ideas and movements from above, things happen which are frequently unexpected and sometimes truly marvellous. Those who are made special instruments in this unfoldment of the Divine Plan can do things which would not otherwise be possible and, being unaware of this fact, are inclined to take credit for what they have done and feel proud of it. But the fact remains that most of what they have done is due to the influx of power, ability and direction from above and not to their individual capacity except in a minor degree.

This is not to deny the credit to these individuals who have done this work and made these expressions possible. They are chosen as the instruments of the plan because they are the best instruments available for this purpose, but the broad plan comes from above, the knowledge and power come from above, the inspiration and direction come from above, and but for these, their contribution to the work in the particular field could not be what it has been.

Those who have contact with the inner planes of existence, contact which may be conscious or unconscious, are aware of these facts and are therefore extremely humble. Those who are not, frequently suffer from the illusion of greatness and take credit which is really not due to them. Let us not grudge them this little satisfaction if they have served as satisfactory instruments of the Plan and made its fulfilment possible.

Nothing exemplifies this descent of a plan from above and its working out in the lower worlds in different stages through a succession of workers, more than the development of the Theory of Relativity. One can almost see with wonder how the ground for the development and formulation of the theory was prepared in the most systematic manner by different workers working in the fields of physics, chemistry and mathematics, all over the world, during the course of a century. And when all the material was ready and the preparation had been completed, here comes Einstein to fuse together all the ideas and discoveries into a wonderful concept which has not only stimulated in a remarkable manner the growth and integration of scientific knowledge but has led to discoveries which may have a tremendous influence on the life and evolution of humanity and may alter the very direction and nature of our civilization for centuries to come.

This is not the place to discuss Einstein's Theory even briefly in order to give the student of Occultism an idea of what it is about. Many small books meant to give the layman a general idea of the theory in simple language are available in the market. It is not necessary for the student of Occultism to acquire a thorough knowledge of this theory, which he cannot do without having considerable knowledge of higher mathematics and modern Science. But it will be useful for him to have a broad general idea of the theory so that he can see for himself how the recent developments in modern Science are gradually undermining the materialistic philosophy and paving the way for an earnest and careful examination of the Occult Doctrine by the modern scientist. He may not as yet be able to persuade the orthodox scientists to examine with an open mind the Occult philosophy but he will certainly be able to find in recent scientific developments much to illuminate the Occult Doctrine. And above all he will have some idea of the kind of fantastic world into which the modern scientist invites him to enter in place of the world of Reality which according to Occultism is hidden within the universe and also within the heart of each human being.

The scientists, as has been pointed out above, are in many cases the instruments of the Divine Plan and have to work whole-heartedly on these scientific problems with the utmost concentration of purpose in order to be able to solve them. Perhaps this would not be possible if they did not have this kind of fanatical faith in the limited materialistic aims of modern science and the larger and more reasonable and realistic world of the Occult philosophy was not blotted out from their mind for the time being. It happens frequently in the evolution of individuals and of humanity as a whole that those who are passing through a particular phase in their evolution or are serving as chosen instruments of the Divine Plan, have their heart so set on the work which they have to do that their mind gets conditioned to a great extent for their work for the time being, and becomes incapable of seeing anything beyond their circumscribed field of work. This is, of course, only a temporary phase in the long life of the soul and lasts only so long as extreme concentration on the particular work in that life is necessary. But there is no compelling reason why others who are not engaged in that kind of work, or even those who are, should allow their mind and sense of discrimination to be clouded in this manner. It is possible for an intelligent and mature individual to have a very wide and comprehensive outlook and open mind and yet to be able to work in a very narrow and limited field, if the plan requires this. But an ordinary man cannot do this.

In considering the important principle of Relativity from the Occult point of view, it is desirable to know to some extent the limited nature of Einstein's theory, its validity, and the kind of mental prospect which is in store for those who adopt the materialistic philosophy sponsored by this theory. This kind of appraisal of the theory will enable the student of Occultism to judge for himself the comparative validity of the two philosophies and their relative value in promoting the true evolution and happiness of man both as an individual and *en masse*.

In order to be able to understand the comparative validity of scientific theories and Occult doctrines we have to be clear in our mind about the manner in which knowledge in the two cases is arrived at. The scientific theories as has been pointed out in the Preface are based on observation of facts, making assumptions of postulates and reasoning from these two to arrive at definite conclusions. These conclusions are then tested by experiments as far as possible to see how far the results of the experiments tally with the conclusions. If the conclusions are not at all confirmed by subsequent experiments the theory is either abandoned or suitably modified in such a manner that its conclusions are confirmed at least to a reasonable extent. If and when the conclusions of the theory are confirmed to a reasonable extent by experiments the theory is tentatively accepted for further research and extension of knowledge.

It will be seen, therefore, that the observed facts on which it is based are the only reliable factors in the formulation of a scientific theory. The process of deducing conclusions from the available data even in mathematical deduction is dependent upon the postulates which are laid down. These may or may not be valid and the conclusions drawn on the basis of these may thus be correct or incorrect. Mathematics can be made to prove almost anything provided we lay down suitable postulates, or in other words, make certain unproved assumptions. It happens frequently in scientific investigations that we are faced with certain facts and phenomena and we have to formulate a theory which accounts for these facts satisfactorily. Every such theory has to be based upon certain unproved assumptions and a careful selection of these postulates can generally be made to prove what we have already decided consciously or unconsciously to prove. The trick lies in working backwards from the conclusions to the fundamental assumptions which have to be adopted in order to arrive at the desired conclusions. Then we have a respectable theory which not only explains satisfactorily all the facts which are known but has the hall-mark of mathematical deduction. But its weakness lies in the fact that it has adopted postulates which suited its validation and ignored facts which went against.

This is not the place to discuss in detail the postulates of Einstein's Theory. But it may be pointed out for the sake of illustration that it has assumed the non-existence of super-physical worlds, it has ignored the existence of mind and consciousness as principles independent of the physical world, it has assumed that there are only three dimensions of space and one of time and ignored many other facts of existence which are part of the Occult Doctrine and have been verified again and again by seers, occultists and mystics of all ages. The fundamental error lies of course in assuming that what cannot be detected and demonstrated by means of physical apparatus cannot exist at all.

Now, in every scientific theory we have to base our conclusions on a very limited number of observed facts and the vast or rather unlimited storehouse of other facts still in the realm of the unknown remains in the background, from which the emergence of only one fact is enough to completely wreck the theory. The danger of new facts emerging from the realm of the unknown and wrecking even a well-established theory has been pointed out very emphatically by Einstein himself in his well-known statement:

"No amount of experimentation can ever prove me right, a single experiment may at any time prove me wrong." —Einstein.

So, we can see from what has been said above both the weakness and the strength of the scientific method in the discovery of truth. Its weakness lies in the insufficiency of facts observed and the consequent incompleteness of the data upon which conclusions are based. Its other source of weakness lies, as has already been pointed out, in the uncertain nature of the postulates which have sometimes to be laid down in the very beginning as a basis of reasoning. There is very little possibility of error in the actual reasoning in top-class work. Its strength lies in the fact that its facts are facts of observation and we can be sure of them as far as they go. Also, the facts, though they are limited in number, can be verified in an objective manner by means of physical apparatus or by mathematical calculations. So, while Science is advancing at a phenomenal rate as far as its empirical knowledge and its applications are concerned the background of its theoretical philosophy is strewn with the wreckage of ever-changing, ever-shifting theories. And there is no hope or possibility of its ever getting out of this situation and knowing truth with certainty, as long as it sticks to its present methods. Facts can always emerge from the dark background of the unknown and wreck its cherished theories.

Let us now consider for comparison the Occult method of arriving at the vital truths of existence concerning man, God and the universe. This method is based upon the inclusion of totality of facts in arriving at the truth, so no fact can emerge from the dark background of the unknown to invalidate all the knowledge we have laboriously gathered. This may appear as a very tall claim for Occultism but let the student study carefully this method before rejecting it on account of its apparent absurdity. This question has been discussed thoroughly in books like *The Science of Yoga* and *Man, God and the Universe* and we need not go into it here. It may only be mentioned that it is possible to know the totality of facts because all truths underlying every aspect of existence, gradually and step by step, integrate into a single ultimate Truth as we penetrate into the inner recesses of our mind and consciousness, and so it is possible not only to know all that is to be known about a particular truth but also to know it in its essence and in the correct perspective by perceiving it against the background of the Whole, the ultimate Truth of existence. That is why all those Seers who have seen the Truth of truths exhort the aspirant not to wander endlessly in the by-ways of partial knowledge but strive "to know THAT knowing which everything is known".

In considering the validity of the Occult doctrines it should be understood clearly that they are not based, as is commonly supposed by sceptics, on speculative thought born of wishful thinking. They are based on facts of experience obtained by direct unobscured and unobstructed perception of the truth in our consciousness, the only reliable and unerring method of knowing anything in the true sense. There are no uncertain assumptions which may prove wrong, no unknown facts which may appear from the realm of the unknown, no reasoning which may be vitiated by fallacy. The truth is seen in its fullness and essential nature by direct perception in one's own consciousness in the state of *Samādhi*.

Since the truths concerning the fundamental problems of life, in which every human being ought to be vitally interested, are spiritual and not material and can be known only by direct perception in one's consciousness, the paraphernalia of physical apparatus or mathematical deduction is not only not necessary but cannot be demanded for proof or demonstration. Even the sceptic should be able to see that it would be absurd to demand that facts concerning our mental and spiritual nature should be demonstrated by means of a physical experiment. Facts of a mental nature can be seen and demonstrated only by mental methods and those concerning our spiritual nature by transcending the mind and becoming aware of our spiritual nature. Like perceives like.

Every individual who wants to know the truths of the inner life directly and not second hand must pay the price for this and develop his own perceptive faculties to be able to perceive them within his own consciousness. If he cannot do this and is not prepared to undergo the long course of rigorous self-discipline which is necessary for this, then he has to remain satisfied with the evidence of the Adepts of

Occultism who have trodden this path of inner unfoldment and found that ultimate Truth in which all other truths are contained in their essential nature. This is not an unreasonable demand. Even in the field of Science one must first develop the capacity to verify for himself the deeper laws and facts of Nature or take them on the word of those who are experts in their respective fields of investigation.

It should not be supposed, however, that we have to depend upon a single mystic, occultist or religious teacher of antiquity for this evidence for the truths of the Occult Doctrine, as in the case of many revealed religions. There are always present in the world, though unknown to the general public, seers, occultists and mystics who have investigated and verified these truths in their own experience to a greater or lesser degree. And though the lines of investigation and areas of experience of every investigator are individually unique to some extent, the realizations of all of them are basically the same. For, they perceive the same eternal truths directly through a spiritual faculty which unlike the mind and the senses does not distort the vision. So, true occult knowledge is as much based on experience and reliable evidence of experts as scientific knowledge. Why should we take it for granted that observations made only through physical apparatus are reliable and evidence only of modern scientists is admissible in the discovery of truth?

What has been said above should not be allowed to create a wrong impression in the mind of the student concerning Einstein's Theory of Relativity. Though this theory is based on a very limited conception of the universe as a purely physical phenomenon and its conclusions can therefore, at best, be valid only in the very limited sphere of physical phenomena, still, it has made a tremendous contribution not only to scientific thought but also in the realm of religion and philosophy. No development in the field of Science has perhaps provided such a rich harvest of fruitful ideas and corroborations of the Occult Doctrine and thus paved the way for the ultimate acceptance of this doctrine by the scientific world.

The most important significance of Einstein's Theory for the student of Occultism does not, however, lie in those conclusions and discoveries which, like the discovery of atomic energy, have revolutionized our material life. Its real importance for him lies in the fact that it has started the process of mental integration of the various principles and forces which underlie the physical universe and which alone Science recognizes at present. The attempt made in this theory to show that gravitation, electro-magnetism etc. are all different aspects of a single force or principle is a tremendous step in the direction of synthesizing the many *tattvas* or principles which underlie the universe into a single principle. If it can be proved that all these forces are derived from a single force or principle in existence, then the way is cleared for extending the idea to mind and consciousness, when these are given a recognized and independent status by Science, instead of being considered merely as a by-product of physical, chemical and biological changes. It will then be much easier to go one step further and show that mind, consciousness and matter are derived from only One Reality—the highest doctrine of Occultism and the ultimate truth of existence.

Having discussed the comparative value of the methods by which truth is arrived at in Science and Occultism and their relative validity let us now consider briefly the mental prospects which these two Sciences place before a seeker after Truth, or mankind as a whole. It would not have been necessary to question the philosophy of Science if it had confined itself to the discovery of natural laws and facts of the physical world with a view to gain control over Nature's forces and exploit them for the promotion of human welfare. But Science has not confined itself to this limited and legitimate sphere of work. It has claimed the right to pronounce its opinions concerning the highest problems of Religion and Philosophy on the basis of the few superficial facts which it has discovered by means of physical apparatus. It presumes to tell us about the nature and origin of this unfathomable and vast universe on the basis of the sensuous data obtained with the help of microscopes, telescopes and photographic apparatus. It treats man as a purely physical phenomenon and his mind and consciousness as a part of this physical phenomenon. It does not believe that there is any spiritual reality underlying the universe or there is any plan or design behind its working. In short, its philosophy is a challenge to the spiritual view of the universe which occultists, mystics and teachers of the great religions of the world have proclaimed in the most emphatic and unequivocal manner throughout the ages. This challenge is not direct and expressed in so many words but is obviously implied in its ignoring completely the Occult doctrines in its philosophy and work.

In view of this supercilious attitude of the orthodox scientists towards the Occult doctrines I think it is permissible for those who consider these doctrines of vital importance for human evolution and happiness, to give to the student a glimpse of the prospect which scientific philosophy offers to the common man and humanity as a whole. I think it would be quite appropriate if we let a scientist himself describe in his own words this dismal prospect which the spectacular development of theoretical Science offers to mankind. Let me quote a few passages from the excellent book by Mr. Lincoln Barnett—*The Universe and Dr. Einstein*—to illustrate the kind of illumination which Science has provided for us in our effort to understand the nature of the universe and the mental prospect which is to draw us and inspire us in the highly technological age of the future. It is towards this El Dorado of intellectual incomprehensibility and spiritual vacuity that we are bound in our search for truth by the methods of modern Science and from which we should try to extract whatever comfort and inspiration we can for our inner life. There could be no more frightening and repulsive prospect than this and we could not find a better means of keeping the common man engrossed in the superficial and meaningless pleasures and pursuits of the physical world. For, who would like to think of what lies beneath and beyond the surface of life with this prospect in its background? One wonders sometimes whether the scientists themselves believe in all these ideas or whether they are mere playthings of the intellect.

The nature of the physical universe, which is the only reality there is according to the arbitrary assumptions made by scientists, is described very neatly and conscisely in the following manner on page 104 of the above-mentioned book:

" A soap-bubble with corrugations on its surface is perhaps the best representation, in terms of simple and familiar materials, of the new universe revealed to us by the Theory of Relativity. The universe is not the interior of the soap-bubble but its surface, and we must always remember that while the surface of the soap-bubble has only two dimensions, the universe bubble has four—three dimensions of space and one of time. And the substance out of which this bubble is blown, the soap-film, is empty space welded onto empty time."

If this is the conclusion concerning the nature of the universe to which we are led as a result of the assumptions we have made in formulating our scientific theories, is it not advisable to re-examine with an open mind the premises upon which our reasoning is based?

It is a good sign that the less orthodox scientists are now gradually realizing that the solutions which they have found to their philosophical problems are not only incomprehensible but bordering on the absurd, and are consequently feeling frustrated. Let us take another extract on page 126 from the same book to illustrate this state of mental confusion and frustration:

"In the evolution of scientific thought, one fact has become impressively clear: there is no mystery of the physical world which does not point to a mystery beyond itself. All highroads of the intellect, all byways of theory and conjecture lead ultimately to an abyss that human ingenuity can never span. For man is enchained by the very condition of his being, his finiteness and involvement in nature. The further he extends his horizons, the more vividly he recognizes the fact that, as the physicist Niels Bohr puts it, 'we are both spectators and actors in the great drama of existence.' Man is thus his own greatest mystery. He does not understand the vast veiled universe into which he has been cast for the reason that he does not understand himself. He comprehends but little of his organic processes and even less of his unique capacity to perceive the world about him, to reason and to dream. Least of all does he understand his noblest and most mysterious faculty: the ability to transcend himself and perceive himself in the act of perception."

It may be pertinent to ask a few questions which arise naturally after reading this quotation. "If all the highroads of the intellect, all the byways of theory and conjecture lead ultimately to an abyss that human ingenuity can never span," why not give up this suicidal course and try that royal road which all seers, occultists and spiritual teachers of the world have trodden to solve successfully the riddle of the universe? "If man is his own greatest mystery and does not understand the vast veiled universe in which he has been cast for the reason that he does not understand himself," why not try first to understand yourself before trying to understand everything else? This is exactly the method of Occultism which has been followed by every seeker after Truth throughout the ages. And is not, on your own showing, "the ability to transcend himself and perceive himself in the act of perception" merely another way of saying that man has to penetrate into his own mind and consciousness to solve the riddle of the universe and the human soul? In short, he has to follow the well-known and tested methods of Yoga.

It is interesting to note how the solutions of the problems referred to in the above quotation are contained in the very quotation itself and yet cannot be seen. This is what happens when the mind becomes conditioned by prejudice or preconceived ideas.

CHAPTER II

RELATIVITY—TOTAL AND PARTIAL—II

The Absolute and Relativity

It was pointed out in the first chapter on this subject that Einstein's Theory of Relativity deals with relativity only in its physical aspect—with the surface of the total universe in existence—and so it must be very limited in its scope from the very nature of things. If there are subtler worlds and subtler aspects of the universe hidden within the physical universe, if mind and consciousness are independent realities of existence and the basis of the universe, then there must be larger and more comprehensive aspects of Relativity, concerned with and embracing all these subtler worlds and their complex phenomena.

It has been shown in *Man, God and the Universe*[1] that a circle with an infinite number of radii and concentric circles is a very apt symbol of manifestation, the radii representing the Monads and the concentric circles the planes of existence as shown page on 17.

It will be seen from the above diagram how the Monads become more and more separated in the world of manifestation as their consciousness recedes from the centre and becomes involved in grosser forms of matter and less subtle states of mind. On the other hand, the more the consciousness of a particular Monad draws along its radius towards the common

[1] By the author, The Theosophical Publishing House, Wheaton, Ill.,USA.

centre, the more it approaches the consciousness of the other Monads and the Reality from which they are derived.

This relation holds true not only between the differentiated aspects of the Divine Principle which a Monad represents

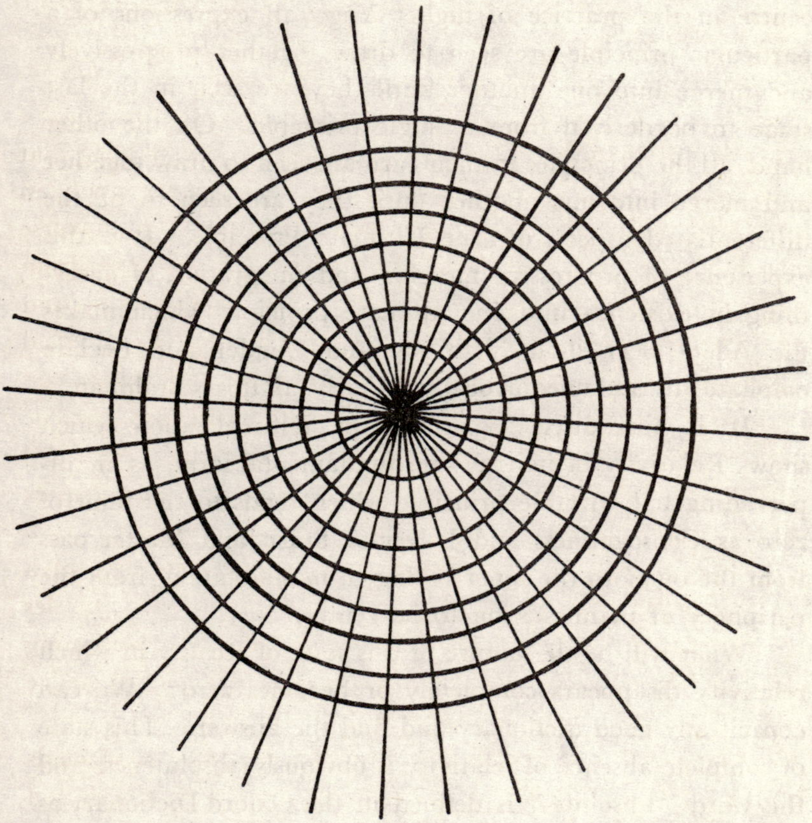

SYMBOL OF MANIFESTATION

and embodies but between the differentiated aspects of any principle which has its source in the Absolute and descends step by step and finds expression in the increasingly dense states of matter. The more it recedes from its central source the more its expressions differ from one another and become more difficult to identify.

These ideas are based not on speculation but on facts of experience. As consciousness recedes towards its common centre in the practice of higher Yoga all expressions of a particular principle are seen to draw together progressively and merge into one another until they are seen in the last stage to be derived from a single principle. On the other hand, all the principles in their turn are seen to draw together and merge into one another until they are seen to be the differentiated aspects of One Ultimate Principle. It is this experience of progressive merging and integration of everything in existence into one ultimate principle which makes the Adept Seer shout with joy and wonder and declare emphatically and unequivocally " Verily, all this is Brahman ".

It is this direct experience in Self-realization which shows Relativity in its true and fundamental form, as an all-pervading but variable principle which tends to the limit of zero as Consciousness and Power or Spirit and Matter pass from the outer to the inner realms of manifestation, from the periphery of manifestation to its centre.

What will be the nature of this state of Reality in which relativity disappears completely or becomes zero? We can consult any good dictionary and find the answer. This state of complete absence of relativity is obviously absoluteness and the word " absolute " is defined in the Oxford Dictionary as " complete, perfect, pure, real, unqualified, self-existent, unrestricted, independent, not relative ". So it is only in the Absolute that there is no Relativity. All else below the Absolute is governed by Relativity, though relativity of varying degrees.

It will be seen therefore that the concept of the Absolute will have a very important and fundamental place in the total conception of Relativity. There may be no place for absoluteness in the very limited and empirical conception of Relativity put forward by Einstein but there is bound to be absoluteness in the Occult conception which includes everything that can exist both in the manifest and the unmanifest. Just as we cannot draw or even conceive a circle without a centre so we cannot have a true conception of Relativity in its total aspect without the Absolute. This conception must include the conception of the Absolute not only because the two conceptions —of the Absolute and the Relative—are complementary and inseparably bound with one another but also because such a conception must include in its consideration the totality of all facts of existence. And it need hardly be said that the Absolute is the most important and final fact of existence beyond and besides which no fact of existence can possibly exist.

This conception of Relativity must include not only the totality of facts of existence but these facts of existence must be based upon reliable experience and not merely upon intellectual speculation or inference. There is finality and certainty only in direct knowledge gained in the realm of consciousness after the mind has been eliminated and transcended. It is only such perceptive experience in consciousness which can embrace the totality of facts of existence because only consciousness can expand to infinity and embrace everything, and not mind. Mind, being a differentiated product of consciousness, shines not only by the light of consciousness but being itself finite limits the vision.

The certainty and finality that there is nothing left to be known about a particular reality to be known, can be present only in the realm of consciousness, from the very nature of pure consciousness and the manner in which consciousness can know a fact of existence by becoming one with it. It is only by such a direct non-dual experience that one can know that Ultimate Truth of existence " knowing which everything is known and nothing else remains to be known ". Such an experience is possible only to the highest Adepts of Occultism and so the conception regarding the total aspect of Relativity can be had only on the basis of the knowledge given out by these Adepts in the doctrines of Occultism. It is true that this knowledge, as given out to the outer world, is vague, sketchy and incomplete but it is certain and final. This means that it is based upon the total facts of existence and its conclusions though vague in their outline are substantially correct and utterly reliable.

It has been mentioned already that relativity decreases as consciousness recedes from the periphery to its centre. This means that relativity is variable, and starting with a zero value in the Absolute, increases progressively as it descends into the lower levels of manifestation, reaching its maximum in the physical world as far as we can see with our limitations. What accounts for this progressive increase in relativity? In other words, what are the factors which bring about this variation? This is a difficult question to answer as our knowledge of the conditions in the subtler realms of mind and consciousness is very vague and incomplete, but still, we may consider a few ideas which will throw some light on the question why and how this progressive increase in relativity is brought about.

The first point to note in this connection is that manifestation is a process of repeated projection, expression and reflection of the One and the same Reality into lower and lower states of itself. As there is, and can be nothing else in existence except the One Reality, this projection, expression and reflection can mean only that at each step the Reality projects itself into a lower level and then expresses itself through this level and in this process of expression reflects to some extent its orginal nature. This process can be repeated over and over again, and thus are formed the different planes of existence and states of mind and consciousness. That is how, in spite of there being only one Reality in existence, there are an infinite variety of expressions at different levels. It is also this fact which enables the Yogi to rise from one level of consciousness to another until he reaches its source, the Ultimate Reality.

Not only is there this repeated projection and reflection of the One Reality, but at each step we have also a further differentiation of the *tattvas* or fundamental principles underlying manifestation. So, we have progressive extension of the process of manifestation both in depth as well as in area, as it were. This is not unlike the spreading out of the spectrum and the separation of its constituents as the dispersed light travels further and further from its source in the prism.

Now, the important thing to note in this progressive externalisation and extension of states of being is that at each step taken, either in depth or in area, relativity is increased and new aspects of it come into being. For, there is one kind of relativity between two states at different levels and another kind of relativity between two states at the same level. Let us not go further into this rather abstruse question here, but the student should be able to see how the principle of relativity becomes far more complex and comprehensive when we try to take a total view of the problem instead of the limited view put forward in Einstein's Theory.

One important advantage of this larger conception of Relativity is that it gives us a standard of reference and judgment with regard to everything in which we have to determine the relative values of two things in respect of any particular quality or feature of life. One of the most disconcerting and demoralizing aspects of modern thought, essentially based on a materialistic philosophy, is that there are practically no natural standards of reference and judgment with regard to anything. There are standards of reference and judgment in galore but they are arbitrary, the chance products of circumstances and whims of prevalent fashions, and not based on any facts or laws of Nature or recognized eternal principles. Although this has been true always, to a certain extent, it is particularly true and dangerous these days when freedom has come rather suddenly, and people having lost even their unsatisfactory traditional moorings, are free to adopt practically anything in their private life right from a dress or style of hair to a mode of thought or standard in art.

The idea that there is, and can be no standard of reference and judgment with regard to these things has perhaps been given a fillip by the Theory of Relativity. It is true that Einstein, in order to develop his theory, found it necessary to consider only the relativity of motion and the impossibility of finding any standard of reference by which to determine absolute motion. According to him all motion was relative because there is nothing fixed in the universe against which the motion of a body could be measured in absolute terms. But though Einstein considered only a very limited aspect of Relativity, the idea which he put forward has produced repercussions in almost every sphere of thought and action and more and more people are beginning to take it for granted that there is no standard of judgment available for anything and therefore everyone is free to do what he likes, provided he can manage to avoid its unpleasant natural or social consequences. There are no real moral values because there is no way of judging moral excellence, or even right and wrong. Everyone can choose his own code of conduct, or do without one, and the only thing to do is to take care that the long arm of law cannot reach him. There is no true greatness because there is no means of judging the quality of human life. Any expert or leader who manages to catch the imagination of the fickle public can become " great " for the time being and is thrown aside on the scrap heap of forgotten personalities when a new individual catches the imagination of the people. There is no standard of excellence in art because people are losing their innate sense of values and perception of beauty and harmony. Ugliness or lack of harmony is extolled as art of the highest class under the cover of high-sounding names and people are prepared to believe anything in these matters if the critics recommend it and it has become the prevailing fashion.

I have put these things in a rather exaggerated form to bring out the undesirable nature of these trends but anyone who observes carefully the chaotic conditions towards which we are gradually drifting will see the danger which is inherent in such conditions. But it is not necessary to enlarge this point further. It is easy to see what kind of effect this lack of any standards in a " free " society is having on our life and what trends it is setting in motion. It is hoped that new standards will be gradually evolved after experimentation with this kind of " free " life but the question is whether such standards will be based on any principles or laws of Nature or be the ever-changing products of passing whims and fashions of the day or of the desire for more and more pleasure and comfort.

The important question we have to consider is: Is there or can there be any natural, reliable and permanent standard of reference in relation to which everything in life can be judged truly and in the correct perspective? Is there anything in this vast ocean of relativity against which we can judge the true values of things in all spheres of life and know reliably in which direction we have to proceed in order to achieve them? The answer to this question is obvious. In a world of relativity only the Absolute and the realities centred round this Ultimate Reality can be such a standard of reference against which anything in manifestation can be seen in the correct perspective and judged in a reliable manner with regard to its value or essentiality.

This standard of reference may not be available to the common man but the important thing to note is that such a standard exists and is available at least to those who have penetrated into the innermost realities of existence and have seen the Ultimate Truth which lies at the basis of the manifested universe. These Seers and Adepts who are the real guides of humanity throughout the ages have tried to communicate to us as best as they can these fundamental truths of existence derived from the Ultimate Truth and shown us how these eternal truths of life and existence can provide permanent, absolute and reliable standards of reference in everything. They have done this either through the revealed religions founded by the great religious Teachers of the world or through the Occult doctrines which have been made public from time to time. A careful study of these with an open mind will give the earnest student at least a general idea of the nature of these spiritual truths and how these truths can provide reliable and permanent standards of reference and judgment not only with regard to our human problems but also the larger problems of Philosophy and Science.

So, let not the sceptic dismiss the idea as purely academic or visionary because the common man cannot have direct access to these innermost truths of existence which centre round the Absolute. Direct perception of these spiritual truths may not be available to the common man but the broad outline of the Occult Doctrine which is based on such perception is available and can be made a standard of reference in his life. If he adopts it as a broad basis of life as an individual or community he cannot go very far wrong even though he may sometimes feel confused and stumble owing to insufficiency of knowledge or will-power. If he ignores it and prefers to live a "free" life without any restraints or direction he does it to his own disadvantage and will have to come back to it sooner or later at the cost of much suffering and loss of time. In a world governed by natural and spiritual laws in every sphere and directed by an overall Plan and Purpose we cannot live lawlessly and purposelessly.

The approximate and vague conception of the spiritual truths of existence provided by the revealed religions of the world and the code of conduct based on them are adequate for man in the early stages of his spiritual evolution but not for the seeker after Truth who wants to free himself from the illusions and limitations of the life on the lower planes of existence. For such a seeker the standard of reference and judgment must be within himself, so that with its help he can traverse safely and effectively the inner realms of mind and consciousness and reach his goal. When he has reached his goal of Self-realization such a standard is no longer necessary for he has become the Truth and so knows It directly.

As the question of the Absolute being the ultimate standard of reference and judgment in a world of Relativity is of great importance, let us dwell on it a little longer and consider briefly its wider cosmological aspects. Coming back to the many scientific conceptions of the universe which are based upon the conclusions drawn from purely scientific and limited data we can see that these conclusions are inevitable as long as we take the universe as a vast chaotic mass of galaxies and stars scattered in boundless space and unlimited time with no centre, no origin, no goal, no basis, no standard of reference to determine the position of anything with respect to time and space. But the real fact discovered by the reliable methods of Occultism is that the universe is not at all what Science with its limited knowledge considers it to be. It is a mental phenomenon rooted in consciousness and based on and derived from an Ultimate Reality. It is because it is a mental phenomenon that it can have a centre from which the mental images in the Divine Mind which constitute Cosmic Ideation are projected. And it is the Ultimate Reality functioning from this centre which is the centre of reference and standard of judgment for the universe projected from It.

As long as Science is ignorant of this aspect of the universe or ignores it and insists on regarding it as a fortuitous concourse of physical galaxies and solar systems, it will never be able to see it in its real nature as a cosmos and will have to put up with the mental chaos in which it has landed itself as a result of its preconceived ideas and materialistic outlook. And our life as human beings, being guided and inspired by this materialistic outlook, will continue to be chaotic. It is needless to point out that when you are living in a chaos anything can happen, from being annihilated in an atomic war to being reduced to the level of civilized animals by having no other purpose in life except physical enjoyment and comfort.

I am not saying that any of these things will happen, for there is a Divine Plan and Divine guidance behind the universe and its affairs are not left to the ignorance and immaturities of those who are sent to evolve in it. I am pointing out these things only to emphasize the undesirable nature of the situation into which we have steered ourselves as a result of the misconceived ideas regarding the nature of the universe and human life and the confused and purposeless life which naturally flows from them. It can be seen easily that unless we begin to rethink and use more our discriminative faculty, conditions in the world will continue as they are, or probably get worse before they can get better.

Now, the point which we were considering was that there does exist a central standard of reference and judgment hidden within the universe and also within the heart of each human being. And this is not merely a remote theoretical possibility but a factor of the greatest practical importance in human life. Art can be judged against the beauty and harmony hidden within the heart of every human being, within Nature and even within the apparent and external ugliness of human life. Human greatness can be judged against the perfection of those highly advanced souls who have attained Self-realization and are still living in the world and guiding humanity from behind the scenes. Moral values can be judged against the laws of spiritual life and the law of Divine unfolding expressed in true righteousness. International morality and behaviour can be judged against the spiritual brotherhood of man which means that we all share a common life and are rooted in the Divine Consciousness. The laws of Spiritual life which are derived from the relation of the Absolute to a human being can provide standards of reference and judgment in everything provided we accept the fundamental truth that a Divine Reality lies at the basis of the universe and not an abstract system of mathematical formulae hanging in a void.

How long it will take us to get tired of this chaotic, meaningless and purposeless state of existence and to seek for meaning, purpose and harmony in our life it is difficult to say, but the Spirit in man must assert itself sooner or later and the time must come when the common man will rebel against this meaningless philosophy of life which has been artificially imposed upon our mind and which we have accepted thoughtlessly to our detriment and confusion. A few people in the world have already begun to have doubts and second thoughts with regard to the basic conceptions of scientific materialism. But these individual trends and thoughts must change into a tide of spiritual awakening before scientific materialism can go out of fashion.

Since the Absolute is the common and eternal standard of reference and judgment with regard to everything in the world of manifestation let us try to understand this point a little more clearly by means of a diagram. It has been pointed out above that if we consider the universe as a vast concourse of countless galaxies and solar systems spread out in boundless and infinite space with no beginning, no end, no origin, no source, then inevitably there can be no standard of reference with regard to anything in the universe. But if we consider the phenomenon of manifestation as the outward mental projection of a Reality from a Centre of consciousness then this Centre provides a natural standard of reference with regard to everything which exists in the mental world so projected.

It is impossible to represent infinite, boundless space without a centre by any diagram and the student can see easily that there can be no *locus standi* for anything in such a void. The Void aspect of the Absolute is also boundless, formless and infinite and cannot be represented by any diagram but then there is another aspect which is called the Great Point or *Mahābindu* in Sanskrit and this aspect can be represented by an ideal point which is the centre of an infinite boundless circle as shown in the following diagram:

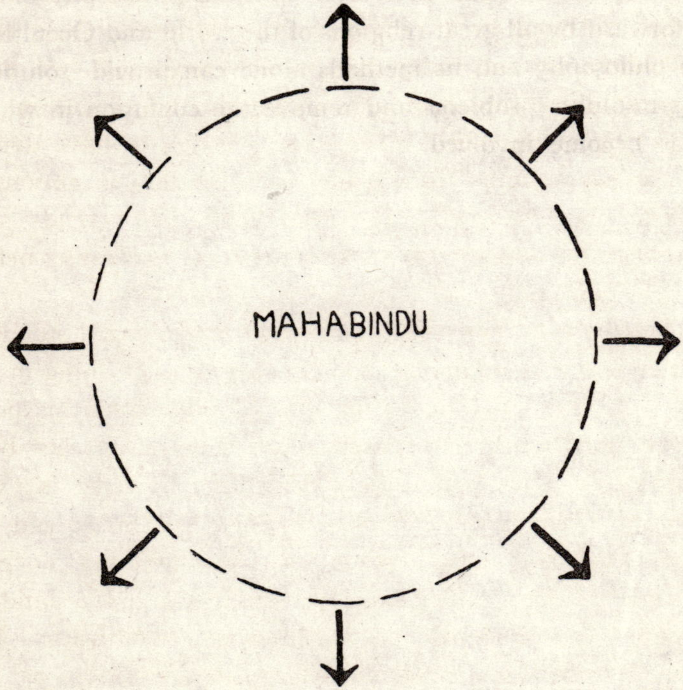

SYMBOL OF THE UNMANIFEST

It is this Great Point or *Mahābindu* in which every reality which exists is centred and from which all that is present in manifestation is projected. And it is this Point containing everything in itself in its essence which can serve as an eternal standard of reference and judgment with regard to everything which exists or happens in manifestation, not only in the present universe but in all the universes which follow one another in the eternal alternation of *Srishṭi* and *Pralaya*.

It will be seen therefore that the cry of indeterminateness and uncertainty in the universe which has been raised by the modern scientific world is due to its very limited outlook and sphere of activity which it has imposed upon itself by rejecting the time-honoured spiritual philosophy of life put forward by all great religions of the world and Occultism. This philosophy and its methods alone can provide solutions to its insoluble problems and remove the confusion in which it has become involved.

CHAPTER III

RELATIVITY—TOTAL AND PARTIAL—III

Relativity and Mathematics

In order that we may be able to gain even a faint glimpse into the nature of Relativity of the most comprehensive nature it is first necessary to consider briefly the relation of Relativity with mathematics. It was pointed out in the Chapter on " Mathematics as the basis of Manifestation " in *Man, God and the Universe* that the relation of all the basic realities in existence is mathematical in nature and it was shown that these realities both in the Unmanifest and the Manifest can be arranged in the order of the basic numerals 0 to 10. But this preliminary treatment of the subject could not do more than show roughly that there is some kind of relation existing between these realities and mathematics. A clearer and fascinating picture of the relation between these realities of existence and mathematics is obtained when we consider Relativity in its total aspect, however imperfect and partial our understanding of this deeper aspect of the subject might be.

Formerly, mathematics was a branch of knowledge cultivated for its own sake and did not find much application in the physical sciences which were still very little developed. But as these sciences developed and it was found necessary to formulate their laws in a more exact manner, mathematics began to find an increasing application in them. Science is realizing more and more the importance of formulating the laws of the physical world in mathematical terms and no formulation is now considered respectable until it has been given a mathematical basis. Not only are these laws sought to be formulated in mathematical terms but mathematics is being increasingly made an instrument for co-ordinating and integrating the phenomena in different fields of Science to discover an underlying unity in the phenomena of Nature. The Theory of Relativity put forward by Einstein is, as has been pointed out in the previous article, a remarkable attempt in this direction.

But though this and other theories have shown the importance and role of mathematics in investigating and formulating the processes and laws of Nature, they cannot bring out clearly and fully the wonderful relation which exists between these and mathematics. In fact, the attempt to investigate and co-ordinate the phenomena in different fields by means of mathematics has raised many new and apparently insoluble problems for the scientists and undermined to some extent their confidence in their methods and theories. They are not now so cock-sure as they were in the first flush of their victories over the forces of Nature, and some of them have begun to feel mentally uncomfortable, though this is not stated explicitly.

The real reason for this confusion which is being created in the field of scientific philosophy lies in the fact that its conception both of manifestation and mathematics is based upon only a very small proportion of the broad facts underlying the universe, facts which are known to Occultism and asserted in the most unequivocal manner by all seers and occultists both in the past and the present. A modern scientist is like an investigator who studies the skin of a fruit and from the observations he has made draws his conclusions with regard to the contents and nature of the whole fruit.

It is only when we consider the totality of manifestation, at least in its broad outline, that we can get a clear and comprehensive though necessarily vague idea of Relativity in its total aspect, and it is only when a faint glimmer of this total aspect is obtained, that the true and wonderful significance of mathematics and its intimate relation with manifestation dawns in our mind. For, Relativity and mathematics are very intimately related and mathematics can be considered merely another aspect of Relativity. They are like the two sides of the same coin, one considering manifestation in its phenomenal aspect and the other in its relational aspect.

It was pointed out in the chapter entitled "Mathematics as the basis of manifestation" in *Man, God and the Universe* that mathematics is the science of pure relations between things devoid of any phenomenal content. When we are dealing with the realities of existence with their contents present we have Nature's phenomena. When we take out the contents or the substance of things which are related, only the steel frame of relations which binds together different aspects and contents of the phenomena is left behind and this is mathematics. It will be seen from this how intimate is the relation between manifestation and mathematics and how our comprehension of both is bound to change *pari passu.* The more our knowledge of manifestation grows and becomes definite the truer and fuller becomes our comprehension of the nature of mathematics and its significance in manifestation. A total view of manifestation is therefore necessary for obtaining a comprehensive and true view of both Relativity and its correlate, mathematics. It is only then that we see the true nature and function of mathematics and are not confused and bewildered by the incomprehensible conclusions to which it sometimes leads.

Take, for instance, the nature of the so-called Godless "reality" at which Science has arrived in investigating the basis of this physical universe. Many modern scientists have adopted the philosophy of materialism and repudiated not only God but also mind and consciousness as independent principles in existence. On the basis of these assumptions they have started to find the reality underlying the universe and discovered that this reality is nothing but a vast and incomprehensible system of mathematical symbols and formulae. "And what the scientist and the philosopher call the world of reality—the colorless, soundless, impalpable cosmos which lies like an iceberg beneath the plane of man's perceptions —is a skeleton structure of symbols." (*The Universe and Dr. Einstein* by Lincoln Barnett.)

This is quite natural because if you try to penetrate into the deepest recesses of matter and refuse to go beyond, you are bound to find nothing but a world of mathematical formulae and expressions which define the innumerable processes and laws of Nature, according to which the machinery of the universe is running. The fact that the fivefold manifested universe, in its subtlest form, is merely a vast system of natural functions is well known to Occultists and has been indicated in aphorism III-45 of the *Yoga-Sūtras*[1] and explained in the following commentary. But while the scientist, owing to his fundamental misconception regarding the nature of the universe considers this world as the ultimate reality underlying the universe, the Occultist "knows" that this world of mathematical formulae and symbols is merely an aspect of the

[1] *The Science of Yoga;* by the author, The Theosophical Publishing House, Adyar, Madras.

Divine Mind and its purpose obviously is to define exactly the Divine functions and natural processes in the manifested universe, so that the laws of Nature can operate with mathematical exactitude. So, in trying to find the ultimate "reality" which is hidden beneath the external universe, while the scientist thinks that he is facing a dead wall which he cannot cross, to the Occultist this dead wall is nothing but a veil behind which are hidden unimaginable worlds of mind, pure consciousness and Reality. It is these worlds into which the consciousness of the Yogi emerges when he is able to transcend the planes of matter and lower mind in the practice of higher Yoga.

It will be seen in considering the relation of manifestation and mathematics that it is not only a question of some correspondences between mathematical truths and realities of manifestation. The two are related to each other as a building and its plan, in which everything in one corresponds to something in the other. In other words, every mathematical truth is matched by some reality of existence in manifestation and every reality in manifestation is represented by some mathematical truth. The manifested universe may thus be considered as a sky-scraper which is supported by a steel frame of mathematics in every part. In constructing a sky-scraper the building follows the outline of the steel frame and what is not provided in the steel frame in a skeleton form cannot form part of the building.

This fact is of great importance in the development of Science and determines what is possible and what is not possible to achieve in its different fields. It is a well known fact that in mathematics only some things can be achieved while others are impossible. Thus we cannot factorize prime numbers. We cannot square the circle. Obviously, what is not possible mathematically cannot find any correspondence in the realm of manifestation, and what is possible, can. On the other hand, it should be possible to express every reality of manifestation by a mathematical expression.

This is, however, only a general principle at which we arrive from a consideration of the relation between manifestation and mathematics. It is not possible for us to say what is possible or not possible with our present limitations. Only those who can raise their consciousness to the level of the Universal Mind and can see clearly and fully this relation between manifestation and mathematics can trace these correspondences in detail and determine these limits of possibility.

But though the ordinary man cannot do this the idea should open a new and interesting field for philosophical thought and scientific research. Scientists are trying to express natural phenomena in mathematical terms. Thus, for example, each colour is now represented by a wavelength which is a mere number. The reverse process of finding correspondences to mathematical expressions in the field of manifestation is also utilized successfully sometimes. The discovery of atomic energy after Einstein had deduced the famous formula $E=mc^2$ mathematically illustrates to some extent this reverse process. The Occultist is in a more favourable position to do this kind of work for his understanding of the process of manifestation is truer and fuller though not so detailed as that of the scientist. It is easier for him to find in the field of manifestation phenomena and natural laws which express or correspond to mathematical truths.

A remarkable similarity and correspondence between manifestation and mathematics is seen when we examine their origins.

It can be seen easily that the four basic operations of arithmetic—addition, subtraction, multiplication and division—are the basis of all mathematics and addition is the primary operation from which the other three are derived. Let us see how this is so.

Subtraction is the result of addition. For, when we are dealing with a self-sufficient system, the apparent addition of something always involves subtraction somewhere. Something cannot come out of nothing, and if it has to come out of something else, it must produce a deficiency in that something. Similarly, if we subtract something from a self-sufficient system we must add an equal amount somewhere else. If we dig a hole in the surface of the earth we must simultaneously produce a heap somewhere else. Subtraction and addition will thus be seen to be complementary and two aspects of one operation which may be called "transposition".

Multiplication is merely mass addition and can be considered as a device for adding together similar quantities *en masse* as shown below:

$6+6+6+6+6=30$; $6\times5=30$.

Similarly, division is merely mass subtraction and is merely a device for subtracting similar quantities *en masse* as shown below. Thus dividing 12 by 4 means merely that we subtract 4 from 12, 3 times as follows $12-4-4-4=0$; as $\frac{12}{4}=3$. Division is thus merely the opposite of multiplication.

From the above considerations we see not only that multiplication and division are two aspects of the same process but also that multiplication is an aspect of addition and division is an aspect of subtraction. This mutual relationship of the four basic operations of mathematics may be represented diagrammatically as follows:

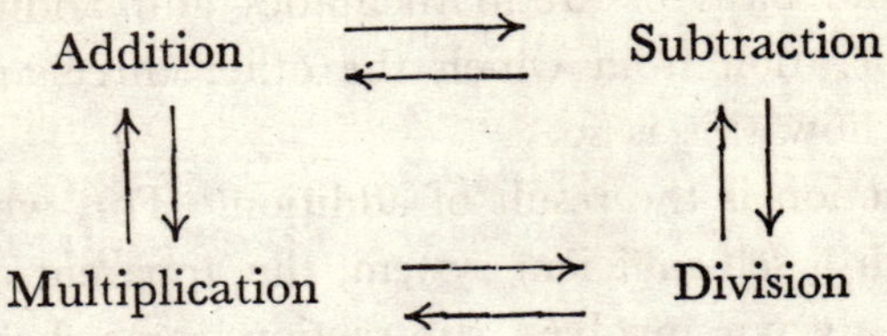

What does the above diagram show? It shows that all the four operations are different aspects and instruments of one single process which may best be called "transformation".

Since mathematics underlies manifestation and the whole process of manifestation from the highest to the lowest levels is, in a way, built upon its foundations and is an image of mathematical truths, it follows that there should be only four basic processes involved in manifestation at the initial stages and these processes should correspond to the four basic operations of mathematics. A close analysis of the Occult doctrine with regard to the essential nature of manifestation, particularly in its initial stages, shows how true this is. In fact, the whole process of manifestation may be considered as the expression of the four basic operations of mathematics, in terms of and through the instrumentality of the various principles or *tattvas* involved in the descent of the Ultimate Reality into progressively denser states of matter and less subtle levels of the mind. Let us see how this relation may be traced.

Let us take first the principle of multiplication. A close consideration of the nature of multiplication will show that it is the principle underlying proliferation. What is proliferation? According to the Oxford Dictionary "proliferate" means "reproduce itself, grow by multiplication of elementary parts, produce thus". In simple multiplication a particular number is multiplied by another and thus grows in multiples. In essence, therefore, multiplication is nothing but the principle of reproduction or proliferation in the abstract.

If we examine the Occult doctrine concerning manifestation we shall see that manifestation, in one of its basic aspects, is nothing but the process of reproduction *ad infinitum* from the highest to the lowest levels. The Ever-Unmanifest Reality multiplies itself *ad infinitum* through the *Mahābindu* or the Great Point and produces countless centres of Reality called Monads which start as atoms, and evolving through various stages in aeons of time, become countless Solar Logoi which are present in the Cosmos. These countless centres of consciousness functioning in manifestation are, according to the Occult doctrine, in their essential nature, reproductions of the Ultimate Reality through the Great Point which exists eternally in the Absolute. This is the highest and the primary stage of manifestation from the Unmanifest and we can see easily that it corresponds to, and is an expression of, the abstract principle of multiplication in mathematics. Are there other expressions of this principle in Nature working at the lower levels of manifestation? Everywhere, every archetype is expressing itself constantly in countless forms. Every natural phenomenon is generally a reproduction of some prototype. If we examine the human body we see every organ conforming to a set pattern, every cell reproducing itself and conforming to a set pattern. The whole working of Nature is a process of reproduction of forms in countless ways and shows the principle of multiplication in action.

What about division? Division, on careful consideration, will be seen to be essentially a process of differentiation. When a whole thing separates into its parts naturally, its parts are related to one another in a definite manner mathematically. Take the case of the division of white light into its constituent coloured lights. We get a regular spectrum in which the colours are related to one another in a series and the wavelengths of the lights of different colours are related mathematically. Differentiation of an integrated state will thus be seen to be the most natural expression of the mathematical principle of division, though division need not always be regular.

If we examine the process of manifestation in its infinite variety we find that it is a process of progressive division into smaller and smaller parts. The *Ashwattha* or the Tree of Life is a symbolic representation of this principle of progressive division permeating all manifestation in every sphere of natural phenomena. Everywhere in manifestation we find a principle or *tattva* dividing itself into smaller and smaller branches as it descends into the lower levels of manifestation. All the principles or *tattvas* are derived from the differentiation of the One Reality and the different aspects of each principle come into play by progressive division.

It should not be difficult to see the mathematical principle of addition with its correlate subtraction finding expression in manifestation everywhere. Manifestation is a flux of phenomena involving change everywhere, all the time. When any change takes place it involves addition and subtraction at the same time. Formation of something new means the disappearance of something already existing. Disappearance of anything means addition of something equivalent somewhere else. Gain of one person means the loss of someone else. We fail to notice this obvious fact because of our limited vision which is confined within a particular area of observation. But we should be able to see at least intellectually that the above principle is of universal application.

We have been considering so far the realm of manifestation in its relation to mathematics. But if we recall what has been said in *Man, God and the Universe* about the Primary and Secondary differentiation of the Ultimate Reality which results in appearance of the *Shiva-Shakti Tattva* on the one hand and that of *Maheshvara-Maheshvarī Tattva* on the other, we shall see in these two eternal Principles in the Ever-Unmanifest the highest expression or rather the source, of the four basic mathematical principles of addition, subtraction, multiplication and division. For, as has been shown in the above-mentioned book, the Primary differentiation of the Ultimate Reality into two polar opposites of Consciousness and Power can be considered merely as a transposition of the same Integrated State into more and less of the Reality, producing the contrast of Consciousness and Power. And the further differentiation of Consciousness and Power into Mind and Nature respectively which is the basis of the Logoic Principle embodies in itself the infinite potentiality for multiplication and division in the realm of the manifest. It is from this Principle in the Ever-unmanifest that an infinite number of universes, Monads and Solar Logoi continue to appear eternally. It is not necessary to go further into this question here but the student should be able to gain from what has been said above some idea of the wonderful manner in which the realities of existence are related to one another both in the realm of the manifest and the unmanifest.

The above truth of the intimate relation between manifestation and mathematics, in which every reality of manifestation has a mathematical counterpart and every truth of mathematics has the potentiality of serving as a basis of some form of manifestation, is of tremendous significance and importance to scientists and students of Occultism alike. It is not surprising that scientists whose knowledge of the nature and process of manifestation is very partial and superficial should have missed the significance of this truth which is related to the Whole. But there is no reason why students of Occultism who have at least a vague and sketchy idea of the totality of manifestation should not have seen this obvious truth.

One point of great significance emerges from the consideration of this relation between manifestation and mathematics and has an important bearing on the validity of the materialistic philosophy which is in fashion these days. The very fact that mathematics is the foundation of the vast and complex physical universe stretched out in time and space suggests naturally to the human mind the mental nature of the universe. For, how else could mathematical knowledge which is a purely mental reality be co-extensive with such non-material and intangible things as time and space and regulate the countless and complex processes involved in the functioning of the universe? It is only if the manifested universe is purely mental that it is possible to conceive its being supported on a framework of a purely mental system of realities like mathematics. It is difficult to conceive a manifested universe which is "a fortuitous concourse of atoms" and therefore functioning in a chaotic manner, built on a purely material basis which is invisible; but considering a Cosmos which is functioning with exactitude and perfection and showing more intelligence, design and plan, both in its microcosmic and macrocosmic aspects, than the most intelligent human being, surely, it is impossible to conceive such a Cosmos without a mental basis of unbelievable intelligence and powers. But our scientists have found incomprehensible realities like time-space continuum, vibrations without any medium in which they can take place, and learned to take such things for granted without understanding them in the least. Obviously, understanding and comprehension are not necessary and cannot be demanded in a system of thought or philosophy which does not recognize the independent existence of mind and consciousness but insists on considering them as by-products of purely material changes. Absurd conclusions can be accepted readily in such a philosophy and the most illuminating and fascinating truths of great philosophical grandeur, verified and attested by countless occultists, seers and great Teachers of world religions, must be rejected.

Even granting that a mental framework of mathematical laws and relations can underlie and pervade a purely physical universe the question arises in whose mind is this vast and complicated system of thought present? For how can there be a vast and complex mental system of thought without a thinker? But perhaps it is quite allowable in the rational world of Science to have thinking without a thinker as it is allowable to have vibrations (of light) without a basic medium like ether. This is another one of those postulates which must be added to the theories of Science in order that God may be effectively kept out of their philosophical thought. Verily, to what lengths has man to go and what incredible things he has to believe when he refuses to believe in God and denies Him in his pride and egoism!

As this is an issue which is very fundamental and relevant to modern scientific thought and research let us dwell on it for a while. We assume in our daily life that when a palatial building is being constructed there must be an engineer who has designed the building and is supervising its construction. The construction of the building must be directed by him at every stage; the plan of the building must be present in his mind. He must have prepared the plan beforehand. The knowledge which enables him to prepare the plan must be present in his mind. If we saw a magnificent building coming up, being constructed systematically with great precision and perfection and we insisted that it had no plan, no designer, no architect, no thinker, no supervisor behind it we would be considered mad.

Now, there can be no doubt that the universe is also based on organized activity and thought with a scope so tremendous that it is difficult for the human intellect to conceive it in its entirety. The presence of evolutionary forces and the exquisitely exact and harmonious co-ordination of all its activities shows not only that there is a plan but also that the plan has been made by a Master Architect and designer. The perfect running and co-ordination of such a stupendous organism of such inconceivable complexity shows that this Supreme Architect must have the most comprehensive and detailed knowledge not only of all the laws of Nature but also of mathematics upon which these laws are based. In whose mind is all this knowledge contained and who applies this knowledge in designing and running the universe? In the mind of insentient atoms and molecules which constitute the raw material of the universe and follow blindly the mechanical laws of physics and chemistry? In the mind of the cells constituting a living body which perform blindly their allotted tasks with such wonderful fidelity and perfection?

Science has not only no answer to such questions but ignores them as if they do not exist or are of no importance to human beings. The whole world of orthodox Science believes implicitly that there is an exquisite and stupendous design in the working of the universe but no designer. There is complete knowledge of the laws of Nature and their varied and mathematical application but no mind in which this knowledge is contained and by which it is applied in the running of the universe. There are evolutionary forces at work in every sphere of life that trend towards greater and greater perfection but no objective and no goal. The universe is moving relentlessly towards a certain death brought about by the exhaustion of the available energy at its disposal. But who created the universe and imparted to it the tremendous amount of energy when the universe began its career? No one! If the universe can be created once and made to run as in the present case why cannot this happen again? And again? Because Science cannot find such a Creator when it looks into its telescopes and microscopes and photographs obtained through these mechanical instruments. This is the kind of logic upon which present modern scientific thought is based and this is the kind of "rational" attitude of which the majority of orthodox scientists are very proud.

Whatever Science and scientists may or may not say on these fundamental and pertinent questions the answer of Occultism and occultists to these questions is quite clear, unequivocal and emphatic. There is an Architect of the Cosmos and He is not only a Designer but it is in His Consciousness that the Plan of the Cosmos is designed, worked out and the universe exists. He is the Architect and the Plan, the mathematician and mathematics, the Dweller and dwelling, the Seer and the seen, all in One. For according to Occultism there is only One Reality from which everything in manifestation is derived and apart and outside of which nothing can exist. This highest doctrine of Occultism is not a mere philosophical concept born of speculative thought but the Truth of truths ultimately discovered by every seeker after Truth by the only method there is of discovering truth in the real sense.

CHAPTER IV

RELATIVITY—TOTAL AND PARTIAL—IV

Relativity and Pairs of Opposites

One of the most fundamental differences between the Occult conception of the universe and the conception of modern Science lies in the degree of complexity attributed to the objective aspect of the universe. Recent researches in the field of physics and chemistry have shown that even the physical world, in which alone Science believes, is far more complex and subtle than had been previously supposed, The discovery of sub-atomic particles and the tremendous amounts of energy hidden within the atoms of matter has laid bare an invisible aspect of physical matter which was not suspected before. It has shown that the range of perception of the realities existing in the total universe depends upon the instruments which are available or which we are able to devise, and if we cannot perceive certain realities with a particular set of instruments it does not mean that they do not exist and cannot be perceived if finer and more appropriate instruments are available. But Science has not learnt anything from these facts and insists that worlds subtler than the physical do not exist because they cannot be detected and perceived by means of physical instruments. And it is not only investigating the phenomena of the physical world on this unjustified assumption but trying to elaborate and place before us grandiose theories with regard to the nature of the vast and mysterious universe, not excluding even its origin and ultimate destiny. Never has man tried to guess so much on the basis of so little.

The Occult conception of the universe, as is well known to students of Occultism, is far more complex and profound and is based on the thorough investigation of the invisible and subtler realms of Nature by the only methods which are applicable to those realms, i.e. by unfolding the inner powers and faculties of mind and consciousness which lie dormant and hidden in every human being but can be made active by appropriate scientific methods of Yoga. This is not the place to discuss the nature of the real and total universe revealed by adopting these methods. These questions have been discussed thoroughly elsewhere in their appropriate contexts. But there is one aspect of this inner universe, of which this physical universe is a mere shell, which we may discuss here on account of its relevance to the problem of relativity. This aspect is concerned with the question of the manner in which an infinite number and variety of worlds can exist, one inside the other, without the inhabitants of one world being aware of the presence of other worlds or their inhabitants.

The existence of these worlds introduces new dimensions in the concept of relativity and shows how limited and partial the scientific concept of relativity is. For, these worlds function in more than three dimensions of space and are governed by variable measures of time. At the junction of the different planes in which these worlds function a new kind of relativity therefore comes into play and has to be considered if we are to have a concept of relativity in its total aspect. And at the basis of all these worlds of different planes lies that Ultimate Reality which is above the realm of relativity.

It was pointed out in an earlier chapter that mathematics provides the steel frame of manifestation and the two are related to each other like an object and its shadow. Since mathematics is the science of pure relations its truths must be applicable to the whole field of manifestation which comes under the domain of relativity. It is, therefore, impossible to understand the significance of its truths in their entirety without having a detailed and correct knowledge of all fields of manifestation existing at all levels. It is possible that we may have discovered a mathematical truth which is applicable only to fields of manifestation which are unknown to us and has no relevance to the facts which are known to us. A forcible application of this truth to facts which are within our knowledge will be unjustified and will naturally lead to wrong interpretation and incorrect conclusions. Many mathematical truths and relations have been discovered by modern scientists which appear to be enigmatic and incomprehensible. May it not be that they represent some truths pertaining to the subtler planes of existence with which we are not yet familiar or cannot fully understand, owing to the limitations of the intellect? Only those who have knowledge of all the truths of existence in their totality, embodied in the One underlying Truth of existence, can therefore work out with certainty and confidence these correspondences between truths of mathematics and those of manifestation.

The concepts of infinity and infinitesimal are two such truths which are the most enigmatic problems of philosophy and the most mysterious entities in the realm of mathematics. Before dealing with these concepts in relation to some particular realities of manifestation let us first consider the purely mathematical aspect as embodied in numbers representing increasing and decreasing quantities. Let us start with 1 and represent its progressive increase in magnitude by putting on the right side an unlimited number of zeros after it, and the progressive decrease in magnitude by putting an unlimited number of zeros on the left side with a decimal point as shown below.

$$100000000 \longrightarrow \propto$$

$$0 \cdot \longleftarrow 0000001$$

We have in the above symbolical representation of magnitudes by numbers all degrees of magnitudes fully represented. These numbers extending to infinity on both sides are flanked by the ideal zero on one side, representing the ideal limit of the infinitesimals, and the ideal infinity on the other side, representing the ideal limit of the infinites.

If we consider carefully the significance of the above relation existing between zero, infinity and all possible numbers in between, we shall see easily that all the three represent one single continuous process and both infinite and infinitesimal are two aspects of the same process which in one direction is called " decreasing " and in the other direction is called " increasing ". Also, in one direction ideal infinity is the limit and in the other direction the ideal zero is the limit.

Now, it is well known to those who are familiar with higher mathematics that both infinity, represented symbolically as $\propto$ and infinitesimal represented by zero, O, are indeterminate. For, if they were determinate it will mean that they have a limit and what has a limit cannot be considered infinite or infinitesimal by any means. Not only are they limitless and indeterminate but they are also relative. In fact their indeterminacy is due to the fact that they are relative. The relative nature of infinity and zero can best be represented by the following composite series, in which on the right we have a series of infinities following one another and disappearing into the ideal infinity and on the left we have a series of infinitesimals represented by a series of zeros disappearing into the ideal zero.

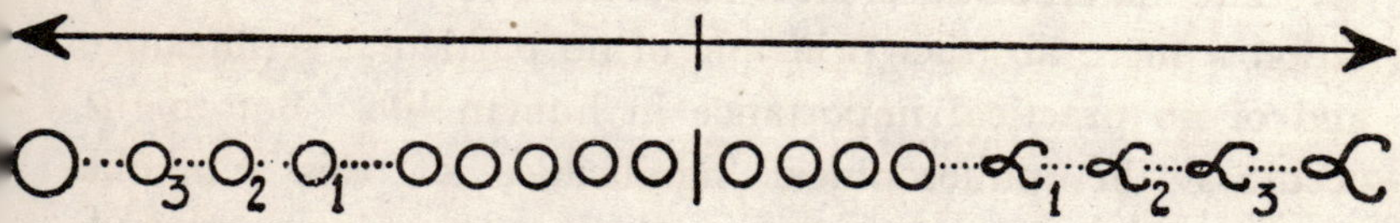

The question now arises: If on one side we have a limitless series of infinities and on the other a limitless series of infinitesimals where do they disappear and if they meet somewhere, how and where? It is here that the new geometry helps us and enables us to gain some faint indication of the mystery of their disappearance into a void. According to the old Euclidean geometry when a straight line is extended in both directions, its two ends go on and on indefinitely, disappear into a void, and do not meet anywhere. But according to the new geometry the two ends describe a cosmic circle of an infinite though indeterminate radius and meet at infinity as shown below:

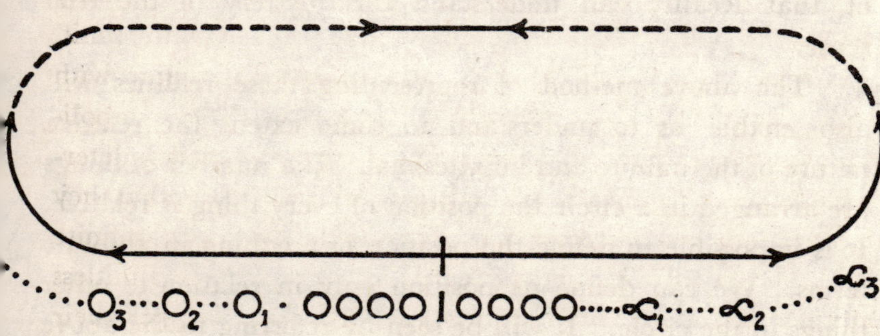

As there are many stages in the increasing infinities on one side and decreasing infinitesimals on the other the use of the word infinity in relation to the meeting ground of the two is not quite appropriate but there is no other way of expressing this merging of the two extremes into one another. Perhaps, the difficulty can be avoided by saying that the two lines meet at ideal infinity.

The materialistic scientist will see in this symbolic representation a mere abstract principle of no particular significance and of no practical importance in human life. But to the occultist the philosophical significance of this symbolic representation is quite clear and easy to understand. It is a well known doctrine of Occultism that at the heart of the universe lies an Ultimate Reality which remains Ever-Unmanifest but is the potential container and solvent of all opposite principles and the origin of everything in manifestation. The infinite and the infinitesimal which are opposites underlying the abstract principle of magnitude must therefore necessarily merge in the Absolute as every pair of opposites does. The great cosmic circle symbolically representing the infinite and infinitesimal in the above figure therefore disappears into and meets in the Ever-Unmanifest in a mysterious manner and only those who can gain a glimpse of that Reality can understand this mystery in the true sense.

The above method of representing these realities will also enable us to understand to some extent the relative nature of the infinite and infinitesimal. If a number of things are arranged in a circle the position of every thing is relative. It is impossible to define the position of anything in absolute terms. We can define its position only in relation to other things in the circle. It will be seen by referring to the above diagram that what we consider as infinite may be infinitesimal in relation to another thing further along the cosmic circle and what we consider as infinitesimal may be infinite in relation to another thing far behind. We shall illustrate this point in detail when dealing with the relative sizes of things but the student should be able to see easily that in the case of a number of things of the same category related to one another, as represented in the above diagram, everything is bound to be relative, depending upon what is before it or behind it.

In our ordinary experiences gained through the sense-organs or perceptions gained through the limited intellect we see only a very small arc of this cosmic circle representing the principle under consideration and within this limited range we can fix standards of judgment and judge things by these standards. But if our perception becomes enlarged and more inclusive our perspective will change and our relative values will become altered completely. Some of the molecules which cannot be seen even by the most powerful microscopes are called " giant molecules " when compared with molecules of ordinary size. On account of the unlimited nature of magnitudes on either side what appears infinite at present may not only appear finite but infinitesimal, and vice versa.

In order that we may be able to understand how infinity faces us on both sides—that of the infinitely large and the infinitely small—let us consider briefly in concrete terms the question of the sizes of particles and wavelengths of vibrations underlying different planes of existence. The relation between the two which enables certain wavelengths to interact with particles of sizes within a certain range of magnitudes may be represented diagrammatically as follows:

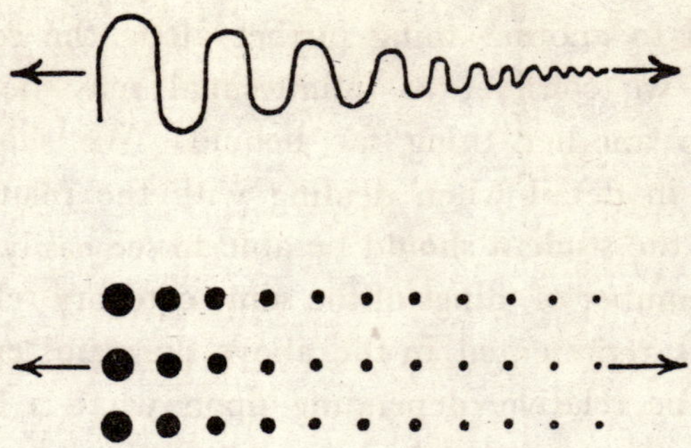

The above diagram is meant to show that the average sizes of particles and average wavelengths of vibrations change *pari passu*. On each plane of existence the band of wavelengths of vibrations operating on the plane corresponds to the range of sizes of particles in such a manner that the two can interact with each other effectively and produce mental phenomena as they do on the physical plane. The whole gamut of these dual realities which determines the underlying form side of manifestation extends on both sides to infinity theoretically and each manifested system occupies a particular place on this infinite scale of particle sizes and wavelengths as shown below:

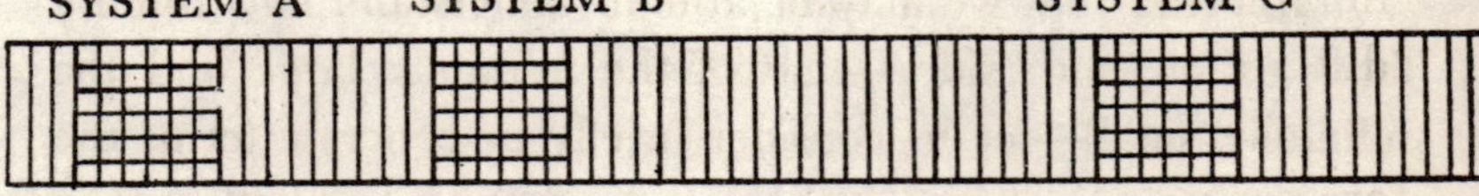

On account of the unlimited extension on both sides there is no limit to size in either direction and therefore theoretically an infinite number of systems with vastly varying sizes of particles and wavelengths can be accommodated in the

Cosmos. The differences between the sizes of particles and wavelengths can be enormous and so one system can remain within another system without any common ground between the two, and therefore without the inhabitants of one system having any awareness of the inhabitants of another system. Thus a system with an average size of particles of 1×10^{-1000} cm (and corresponding wavelengths) will not interfere in any way with a system with the average size of particles on 1×10^{-200} cm as the latter will not interfere with our system. Not only will they not interfere but they will also not be able to come in contact with each other on the material plane and their inhabitants, with their sense-organs organized for their particular range of vibrations, will not be able to cognize the phenomena or inhabitants of the other systems. The experiments made with X-rays illustrate, to some extent, this non-interference of one system with another, but since in this case the vibrations and particles belong to the same system in manifestation these differences cannot be very great and so the non-interference is not complete. The differences between different manifested systems can be comparatively enormous with regard to size of particles, wavelengths of vibrations, etc. and so one system cannot interfere with another in the slightest degree. It is the limitless extension of the principle of magnitude, both in the directions of the infinite and infinitesimal, which makes these enormous differences with regard to these things possible.

The question arises: If different systems in manifestation are separated by such enormous differences and no contact between them is possible on the material plane, how is it possible to establish communication between the different planes and to acquire knowledge with regard to these planes? For, such knowledge is without doubt possible, as is clear from many aphorisms in the *Yoga-Sūtras* (III-50, III-55). In fact, the total knowledge, concerning the universe upon which Occultism is based is obtained by these methods of Yoga.

The answer to the above question lies in the fact that perception takes place in consciousness and is a function of the Seer (II-20). The vehicle, on the other hand, is part of the objective mechanism, the Seen (II-18). The sizes of particles and wavelengths, ete. affect only the objective mechanisms which belong to the differentiated states of the mind. Consciousness, on the other hand, being an integrated state, contains within itself potentially all possible ranges of perception and can therefore respond to all states of existence in which worlds of different ranges of vibrations and sizes of particles function. All that is necessary is that the individual should have developed non-instrumental perception referred to in aphorism III-34 of the *Yoga-Sūtras*. A careful study of the third chapter of the *Yoga-Sūtras* will make this point quite clear.

So, an individual who can rise in *Samādhi* to the required level can be conscious of and function in all these manifested systems. It is difficult to say how high the individual should be able to rise in consciousness in order to be able to embrace other solar systems and planes of being, but some indication of this is given in the *Yoga-Sūtras* in the aphorisms referred to above.

The student should, however, be careful not to confuse manifested systems existing on *different planes of being*, one within the other, (on the particle size-wavelength scale referred to above) with the far-flung solar systems and galaxies on the physical plane with which astronomers deal. The latter are all on the physical plane, on the same plane on which our physical earth exists and in which we function as conscious individuals. They belong to the same range of vibrations and particle size and it is therefore possible to establish communication with them through our sense-organs and instruments like the telescope or microscope. The former, on the other hand, are in entirely different ranges of vibration and particle size and so quite beyond the scope of our physical instruments. They can be reached only from within, through the subtler levels of mind and consciousness.

We have seen above that the ever-increasing magnitudes in the realm of infinites and the ever-decreasing magnitudes in the realm of infinitesimals merge ultimately in the Ultimate Reality in its Superintegrated State. In order to be able to understand how the Absolute is thus the solvent of the two opposite states, it is necessary to have some idea of the integrated state and how this state is related to and finds expression in the differentiated state in manifestation. This question has already been dealt with to some extent in *Man, God and the Universe* and we shall consider it again more thoroughly in a subsequent chapter. Let us consider here only the first step in the progressive differentiation of an integrated state in the form of polar opposites.

Every kind of integrated state, when it is seen in its general aspect, *appears* as a pair of opposites in the realm of manifestation. These pairs of opposites which are seen by the intellect when a principle or *tattva* is polarized by it are called *dvandvas* in Sanskrit, the word *dvandva* here meaning a pair of opposite conditions or qualities. There are innumerable pairs of opposites in Nature which belong to this category and the increasing and decreasing magnitudes including infinites and infinitesimals respectively constitute such a pair. Let us deal with this *dvandva* in some detail so that we can get some general idea of the factors underlying the polarization of principles and their appearance in the form of such pairs of opposites.

In order to understand the nature of these *dvandvas* it is necessary to consider first how a principle is seen in its projection from its integrated state in the unmanifest into its differentiated state in manifestation. All principles are present in their integrated state in the *Mahābindu* or the Great Point. When they are projected from a higher to a lower plane through a centre they may be considered to come out in the form of circles as the waves come out of the centre of disturbance when a stone is thrown into the still water of a pond. As long as our consciousness is centred in the world of Reality and we are aware of its integrated state we can see the whole of the circle, as it were, and therefore there is no *dvandva* produced in our mind. But when we are not at the centre of integrated consciousness but in the realm of the mind and see the particular principle not from the centre 0, as a whole, but from a point A outside the circle, through the instrumentality of the limited intellect, we can see, as it were, only a small arc of this circle at a time as shown in the diagram on the next page.

This arc BC of a cosmic circle which appears as a straight line within the limited range of our vision seems to extend on both sides indefinitely and disappears into the invisible realm of consciousness. The extremities do not come to a definite end and yet they seem to disappear imperceptibly into the intangible void of consciousness in which our mind is embedded. Under these conditions we naturally see the abstract principle as a *dvandva* or polar reality extending on both sides indefinitely.

But the actual fact is that this is merely an illusion created by the mind by seeing the reality of the thing partially and imperfectly in the realm of manifestation through the instrumentality of the intellect. The thing exists really as a whole in an eternal, non-polar form in the world of Reality but only

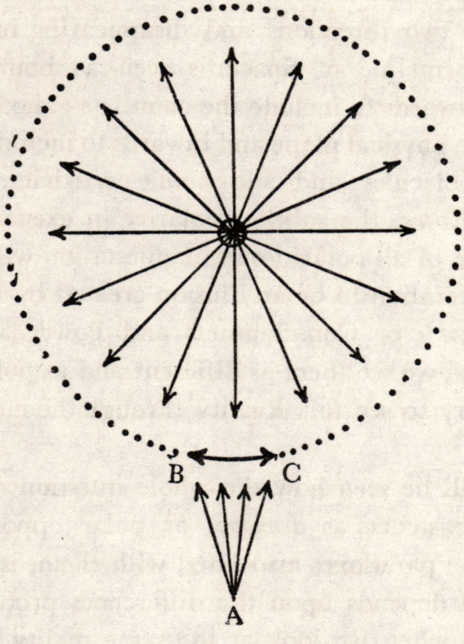

appears as a pair of polar opposites, disappearing into a mental void, because the Absolute can be conceived in the world of the Relative only in this form.

Seen in this light the two polar opposites of the Infinite and the Infinitesimal will be seen to be an illusion produced by the intellect when it sees the principle of magnitude from its limited point of view. The principle itself can be perceived in its real form only from the centre when it is seen as an indivisible whole in the world of Reality. The diagram given above is an attempt to represent this subtle fact symbolically by means of a geometrical figure. What has been said about magnitude holds true of any principle which is seen by the intellect in the world of the Relative. The Principle of Time or Eternal Duration, as it is sometimes called, is seen by the intellect as past and future stretched out limitlessly in two directions and disappearing into a mental void. The principle of Space is seen as boundless space stretching outwards to include the countless galaxies and solar systems on the physical plane and inwards to include the worlds of atoms, molecules and sub-atomic particles. Even the *Shiva-Shakti Tattva*, the subtlest polarity in existence and the hidden source of all polarities in manifestation will be seen on careful consideration to be an illusion created by the intellect. *Shiva* and *Shakti* or Consciousness and Power are one and indivisible and we see them as different and as polar opposites because we try to see this Reality through the medium of the intellect.

So, it will be seen how the whole question of seeing the realities of existence as *dvandvas* or polar opposites by the mind, and the paradoxes associated with them, is a matter of *perception* and depends upon the differences produced in our consciousness when we look at the same reality from within, from the plane of pure consciousness, or from without, from the plane of the mind. In other words, it is a question of relativity, nothing else. There is only One Reality in existence, eternal, changeless, whole, but in the worlds which are present in that Reality we are imprisoned in our respective minds and confined to different spheres of space and time and levels of manifestation. When we look at the realities of existence through our different minds we naturally see them not only differently (IV-15 *Yoga-Sūtras*) but as polar opposites for reasons given above. That is how the intellect is always playing tricks on us and creating mental problems which seem to have no solution. They have not, as long as we are confined within the prison of our mind. They cannot be solved but only dissolved or resolved when we rise above the planes of the mind and the world of Reality dawns in our consciousness. This is the only real "solution" of not only any particular problem but of all problems and paradoxes. There are no problems and paradoxes in the world of Reality.

CHAPTER V

THE DUALITIES IN NATURE

In studying the relation of matter and radiation we find that light or radiation is a dual phenomenon. On the one hand, the phenomena of diffraction and interference show that light *must* consist of waves. On the other hand Einstein's Photo-Electric Law shows that light *must* consist of photons or individual particles of energy. Again, we find that the electrons, the essential constituents of all atoms which determine its properties, show a dual character. On the one hand, they behave like particles having mass and on the other, like waves showing diffraction patterns. Light, even when considered merely as a radiation or wave phenomenon, also shows another type of dual character. It consists of electromagnetic waves, combining in itself both electric and magnetic properties.

A mysterious and remarkable duality pervades all Nature but the instances of this duality referred to above are particularly significant because they have been discovered experimentally by Science unexpectedly and cannot be explained or understood on the basis merely of knowledge available to scientists. Before we consider this duality found everywhere in manifestation and show its origin according to the Occult doctrine, let us first point out a few other instances to show how remarkable and all-pervading it is. The whole field of manifestation is really the meeting ground and the product of two intimately related and inherently opposite principles operating in Nature but it is only where the contrast becomes pronounced that the duality becomes apparent.

In the field of physical phenomena electricity plays a very remarkable and fundamental role because it exists in two opposite forms. Take the atoms which may be considered the foundation stones of the structure of the universe. Each atom according to Science is like a miniature solar system with a central nucleus and a number of electrons revolving round the nucleus at tremendous speeds in circular or elliptical orbits. The nucleus contains a positive electrical charge and each electron carries a negative charge, the total negative charge on all the revolving electrons in the atom being exactly equal to the positive charge on the nucleus. The over-all neutrality of an ordinary atom is thus only apparent and is due to the perfect neutralization or balance of positive and negative electrical forces which exist within the atom. Since the physical and chemical properties of different atoms depend upon the number and arrangement of the electrons and the number of protons and neutrons in the nucleus the infinite number of properties of objects which are cognized by our sense-organs are all based upon and derived from electrical forces in their dual form.

In this perfect balance of positive and negative electrical forces in the atom another kind of duality is also involved—the duality of centripetal and centrifugal forces. For, the electrons which revolve in their orbits with tremendous speeds are held in their respective orbits by the balance of these equal and opposite centrifugal and centripetal forces. The duality of centripetal and centrifugal forces plays a very important role not only in the microcosmic world of the atoms but also in the macrocosmic world of the heavenly bodies. The planets revolving round the sun are held in their orbits permanently by the perfect balance of centripetal and centrifugal forces of gravitation and inertia. In devising the artificial satellites which revolve round the earth Science has merely imitated the Great Architect of the universe who is hidden behind all processes of Nature.

Another kind of duality in the field of physical forces is that of potential and kinetic energy. Energy, as is well-known is of two kinds. When energy is not being spent and no work is being done, energy is said to be potential. When work is being done and energy is being spent it is said to be kinetic. It should be noted that in the instances of dualities referred to above one of the elements of each duality is related to or finds expression in a static state, while the other is related to or expresses itself in a dynamic state. This fact is of great significance as we shall see later on.

Passing from the field of purely physical and chemical phenomena to those of a biological nature in which life forces are involved we again find this duality manifesting itself everywhere. This duality also involves the coming together of forces which are opposed to each other and by their constant adjustment make it possible for living forms to be produced, maintained or removed in their environment. We see male and female bodies coming together for creating living bodies and helping in the perpetuation of the species. We see forces of decay and growth working together in creating, maintaining and destroying these living bodies. One has only to study the working of the human body in detail to see not only the presence of these opposing forces side by side but their constant and wonderful adjustment in maintaining the body in a state of health and efficiency.

Rising from the field of biological phenomena into those of the mind and consciousness we again find this duality manifesting itself everywhere, though on account of the subtler and more complicated nature of the phenomena it is more difficult to detect and define. The primary duality which characterizes the mind in its functioning is that of the subjective and objective states. The mind can either be turned outward and be in contact with the external world objectively or it can be turned inwards and be busy with its own mental images subjectively. When it is functioning objectively it is working through its instruments in the external world, the sense-organs. When it is functioning subjectively it has to depend upon the faculties of memory and imagination for the most part. But though the mind functions in these two ways which are diametrically opposite it is very difficult to separate and distinguish clearly between the two functions clearly. They are so closely intertwined and appear to be bound together by some indissoluble bond.

One thing which obviously binds together the subjective and objective functions, both when the mind is working subjectively or objectively, is perception—the relation of the perceiver and the perceived. Whether the mind is busy with external objects or with its own internal images, perception must be there or the functioning of the mind stops. This is an important point to be noted carefully. A primary duality is always essentially a triplicity in manifestation, the nature of the connecting link depending upon the mechanism through which the duality is functioning. Duality manifests in the field of mind and consciousness in innumerable forms which the student can himself identify by careful examination of mental phenomena.

The above mentioned instances of the existence of duality in physical, biological and mental phenomena should suffice to show the all-pervading nature of dualism underlying all aspects of manifestation. The observation of such a universal phenomenon in Nature points unmistakably to the existence of a dual Principle at the very heart of the universe. Science does not consider that this external universe is the manifestation of an Inner Reality and that the various kinds of phenomena which we find in manifestation are the temporary expressions of eternal Principles which are integral parts of that Reality. So the dualism which Science has found manifesting in physical phenomena is merely a fact of Nature which needs no explanation and no relation to an underlying cause. In fact, the very idea of there being any sequence of cause and effect in the processes of Nature is at a discount. Since the introduction of Quantum Physics, Science has begun to deal with its problems in terms of statistics and probabilities and not in terms of causality and determination, the two basic principles of existence upon which the old Science was based.

But Occultism does not need to change every now and then its conceptions concerning the fundamental principles of existence, for, its knowledge is based not on the shifting sands of limited data collected from the infinite storehouse of unknown facts and reasoned upon the basis of unproved assumptions. Occult knowledge is based upon the direct perception of the fundamental principles and eternal realities of existence in the realm of one's consciousness. This method is not subject to error and uncertainty though the direct knowledge so obtained is not available to everyone and has to be acquired by a long and severe course of self-discipline.

So, let us see what Occultism has to say about the cause or origin of this all-pervading dualism found in nature. The origin of this mysterious dualism is found in the following aphorism:

Shiva-Shaktyātmakam Vishvam

" *Shiva* and *Shakti* are the innermost soul of the universe "

The nature of *Shiva* and *Shakti* and their mutual relationship is one of the most fascinating concepts of the Occult philosophy and it is in this mysterious relation that we have to seek for the origin and resolution of the mystery of dualism pervading the manifested universe. This subject has been discussed in some detail in *Man, God and the Universe*, and it is therefore not necessary to deal with it here again. But a few basic ideas concerning the polar cosmic Principles of Consciousness and Power referred to as *Shiva* and *Shakti* in Hindu philosophical thought may be recalled in order to throw some light on the mystery of dualism which we are considering.

According to the Occult doctrine the Ultimate Reality or the Absolute from which this universe is derived and upon which it is based is a Super-integrated State which contains all fundamental principles and realities which form the framework of the manifested universe in a perfectly harmonized and neutral condition. Such a state in its ultimate aspect must by its very nature be a Void and a Plenum at the same time. Not only this, but in its next lower aspect it must be a pair of two integrated states which are polar opposites and can therefore merge in each other perfectly to produce the Void state of the Absolute. It is these two polar Principles or States which may be considered to be derived from the primary differentiation of the Super-integrated State which are referred to as *Shiva* and *Shakti*.

With regard to the nature of these polar states and their mutual relationship it can be seen easily that such a transcendental Reality must be, by its very nature, beyond the comprehension of the limited and illusion-bound human intellect. But the purpose of philosophy is to enable a human being to understand within the realm of the intellect and with the help of the intellect and intuition these transcendental Realities as far as possible. These Realities, though they are transcendental, are not considered " unknowable " but this knowing by direct perception is possible only in one's own consciousness after undergoing a long and arduous course of self-discipline. To enable an aspirant to tread this Path of Self-unfoldment, those who have gained direct knowledge of these Realities have tried to communicate, as best as they could, some idea of these inner realities through the inadequate medium of the intellect and the still more inadequate medium of language. It is this essential but true and reliable knowledge, partially available to us in the outer world, which is generally referred to as Occultism.

What has Occultism to say about this dual Principle which lies at the very heart of the universe and which is the origin of all the dualities which we find everywhere in the realm of manifestation? A great deal; but we shall merely refer here briefly to a few conceptions which are relevant to our present purpose.

The first important point we have to note carefully is that this duality is polar in its nature and this polarity is inherent in the very nature of a dual Principle derived from a single Principle which appears perfectly neutral and void to the intellect. We see this everywhere even in the field of natural phenomena with which we are familiar.

The second point to be noted is that the two elements of a polarity are opposite in nature. They are generally referred to as positive and negative, but positive and negative are relative terms and we can call either of them as positive and the other negative. Any one of the two poles of a magnet can be called positive and the other negative though we have arbitrarily chosen to call one of them as the positive pole. This opposite and complementary nature of the two Principles of Consciousness and Power has given rise to some ambiguity with regard to the nature of this Reality but the philosophical difficulty disappears if we keep in mind their opposite and complementary nature.

The third point to be noted is that the two elements of a polarity are inseparable, indissolubly bound together and therefore always found together, though one or the other may be predominant in any degree between the two extreme limits of infinity and zero. The whole of manifestation is thus a product and interplay of Consciousness and Power and both are present everywhere in the most intimate relationship. The indissoluble nature of the polar relationship implies the inherent presence of a third element which serves as a link between the two polar opposites. But as we are considering here only the question of dualities we need not go into this question in the present context.

Manifestation is a repeated projection, reflection and expression of the One Reality in progressively lower states. That is why every aspect of Reality which has its origin in the Ultimate Reality is found to be present everywhere and at all the levels, though we may not be able to trace these correspondences on account of the complicated nature of the phenomena and our limited knowledge and vision. But if the fundamental Occult doctrine stated above is true it follows inevitably that every fact of existence must be not only an expression of the One Reality but must contain all fundamental aspects of that Reality within it, whether we can perceive them or not. For, that Reality, in one of its aspects, is indivisible and Whole and the whole of it must be present in every part. This is not an interesting philosophical speculation but a fact of experience at the deepest levels of consciousness.

If manifestation is a repeated projection, reflection and expression of the One Reality in progressively lower states and a dual Principle lies at the very basis of the manifested universe then it is obvious that we should look for the origin of all dualities which we find in natural phenomena in this primary and dual Principle and must expect these dualities to be the expression and reflection of this Principle. It is in this sense that the occult maxim *Shiva-Shaktyātmakam Vishvam* referred to above has to be understood.

Let us therefore see whether we can trace some kind of relationship and similarity between the underlying dual Principle referred to as the *Shiva-Shakti Tattva* and the physical dualities which have been observed by Science in the field of physical phenomena and some other fundamental dualities which are known to us. But before we try to do this let us recall very briefly the respective attributes of the two elements of this primary Duality. As these two elements are opposite and polar in their nature, these attributes enable us to distinguish easily between them.

According to Occultism *Shiva Tattva* is the Principle which may best be referred to as Cosmic Consciousness and *Shakti Tattva* as Cosmic Power. All expressions of Consciousness in manifestation are based upon and derived from the former and all manifestations of Power from the latter. Some of their respective attributes which serve to contrast their polar nature may be arranged in the following manner.

Shiva Tattva	*Shakti Tattva*
Consciousness	Power
Potential	Active
Spirit	Matter

Shiva Tattva	*Shakti Tattva*
Changeless	Constantly changing
Root of subjective phenomena	Root of objective phenomena
Principle of stability	Principle of movement
Source of Will	Source of Action
Positive	Negative
Static	Dynamic

The above table serves to bring out not only their polar nature by contrast but also their complementary nature, for there can be no power without consciousness, no activity without potential energy, no change without a changeless substratum, no action without will, no objective phenomena without a seer to witness them. The two elements of each individual attribute are merely two forms or aspects of the same attribute, one related to Consciousness and the other to Power.

Let us now examine very briefly, in the light of what has been said above, some of the fundamental dualities found in nature by Science or Occultism. It will be found on careful examination of each phenomenon that it consists of two elements and is the joint product of these two elements. One of these elements is related to and derived from the *Shiva Tattva*, the positive and potential element of the Primary Duality and the other from the *Shakti Tattva*, the negative and kinetic element of the same Duality, the two elements being indissolubly bound together in a mysterious combination.

Time, space, matter and energy are the basic realities of the manifested worlds. In describing the physical universe Science uses three quantities: time, distance and mass. Energy is measured in terms of mass, time and distance. The spectacular developments which have taken place in different fields of Science and which have led to the formulation of the Theory of Relativity have shown that time and space are intimately related and indissolubly linked together. This led Einstein to weld them together into one concept in his theory. He called this reality, compounded of two such apparently dissimilar realities as time and space, "time-space continuum". This is one of the incomprehensible concepts of modern Science but it has played a very definite part in the development of the Theory of Relativity and though incomprehensible has been accepted by the scientific world as a mysterious fact of existence.

To the student of Occultism it is of interest because the logic of scientific developments has forced the scientific world to recognize the intimate relation of time and space and weld them together in one concept. Occultism has always recognized and known the intimate relation of time and space and regarded them as two aspects of one reality, though for a different reason. This reason has the virtue, at least, of being comprehensible though it cannot be advanced as a "proof" of the real nature of time and space and the intimate relation which exists between them. This knowledge is possible, as pointed out elsewhere, only when the nature of the *krama* or the mechanism of manifestation is perceived in the light of Reality.

With regard to the third basic unit of measurement used by Science in its measurement and calculations, which is referred to as "mass", recent scientific developments have led to very unexpected and spectacular results. These have completely revolutionized our ideas regarding matter and led to tremendous advances in the field of scientific technology. Here again Science has found a duality in the nature of matter and energy. Matter was measured by its property of mass alone and energy was considered to be associated with mass as an independent reality of existence. But the recent developments in different fields of Science have shown conclusively that mass and energy are interchangeable and these two apparently different realities are only two aspects of one reality which may be called " mass-energy substance ". The nature of this underlying basis of physical reality is again a mystery to scientists and has thrown all their concepts regarding the physical world into utter confusion, out of which there does not appear to be a way.

To the student of Occultism this is a development which was to be expected and is very welcome for two reasons. In the first place it has deprived orthodox Science of its material basis by showing that matter is nothing but condensed or bottled-up energy and thus invalidated scientific materialism. In the second place, it is a great step in the direction of the unification of concepts—the reduction of all matter to elements and then to a few types of particles; the reduction of " forces " to a single concept " energy ", and then the reduction of matter and energy to a single principle which may be referred to as " mass-energy substance ". Occultism has always *known* that the whole universe, visible and invisible, is derived from energy and that energy is derived from mind and consciousness. So, only one more step in the direction of unification of concepts is needed to bring Science to the door of Occultism.

Since these two dualities, Time-Space and Matter-Energy are very important both from the occult and scientific points of view and can provide " bridges " for establishing communication and interchange of ideas between their respective philosophies we shall first consider in some detail the two dualities in the subsequent chapters.

CHAPTER VI

THE DUALITY OF TIME AND SPACE—I

Dimensions of Space

Unlike Science which recognizes only the physical world of three dimensions of space, Occultism asserts that there are also worlds of higher number of dimensions hidden within the physical world. These worlds exist one within the other and can be cognized when man is able to develop the mental and spiritual faculties which are necessary for this purpose. These faculties exist in every human being in an undeveloped state and can be made active by those who are mature enough to function on these subtler planes and are prepared to undergo the rigorous and prolonged self-discipline of Yoga.

A general idea of dimensions of space has already been given in a chapter in the book *Man, God and the Universe* and the student would do well to go through that chapter before entering into the deeper aspects of the subject in this chapter. The Occult ideas regarding higher dimensions of space should not be considered of the nature of speculation. They are based on the experiences and investigations of Adepts of Occultism and even those who are able to develop psychic faculties can have some direct experience of them.

Higher dimensions of space have a recognized place in mathematics. Mathematical analysis can give us not only some general idea of the higher dimensions of space but also enable us to calculate certain properties of objects belonging to worlds of higher number of dimensions. Because objects existing in a world of higher number of dimensions cannot be visualized by those who are confined to a world of lower number of dimensions, conclusions drawn on the basis of mathematical analysis are purely abstract and have therefore to be considered as speculative. But they do show that worlds of higher dimensions of space are possible, and if we recall what was said about the relation of mathematics and manifestation, they are not only possible but must exist somewhere in manifestation. We shall deal with the mathematical aspect of the problem later on and consider some other aspects first to clear the ground.

From what has been said in the chapter on dimensions of space in *Man, God and the Universe* it should be clear that going into worlds of higher number of dimensions means essentially going into greater depths of consciousness. This is a point which we should try to understand clearly if we want to gain some idea regarding the nature of consciousness and also regarding the nature of higher dimensions of space.

In order to be able to grasp the idea of different depths in consciousness let us first consider the nature of depth in relation to some other things with which we are more familiar. Let us first start with a material substance like water. We can understand easily what going into increasing depth of water means. When we dive and enter into progressively greater depths of water in the sea, for example, it is not merely a question of going into a new and ever-changing environment which contains new and unfamiliar things, although this does happen. The deeper levels differ not only in regard to contents but in many other things. For example, as we go deeper, the pressure of water goes on increasing, the light becomes dimmer and so many other kinds of changes which can be measured by physical methods take place. All these changes affect and determine the nature of the flora and fauna found at different levels.

Let us now take another thing like light which is not material. Is it possible to go into greater depths of light? Yes. We all know that when white light is dispersed by a glass prism it breaks up into the form of a spectrum in which lights of different colours and wavelengths are arranged successively in a continuous manner. We also know that the wavelengths decrease as we go from the red end of the spectrum towards the violet end. Of course, there are also invisible parts of the spectrum called ultraviolet and infra-red, which continue on both sides without any limit. They are invisible but they can be detected by their physical and chemical effects.

If we pass a beam of white light through a large-sized glass prism the emergent light will be a spectrum of coloured lights arranged as shown diagrammatically below:

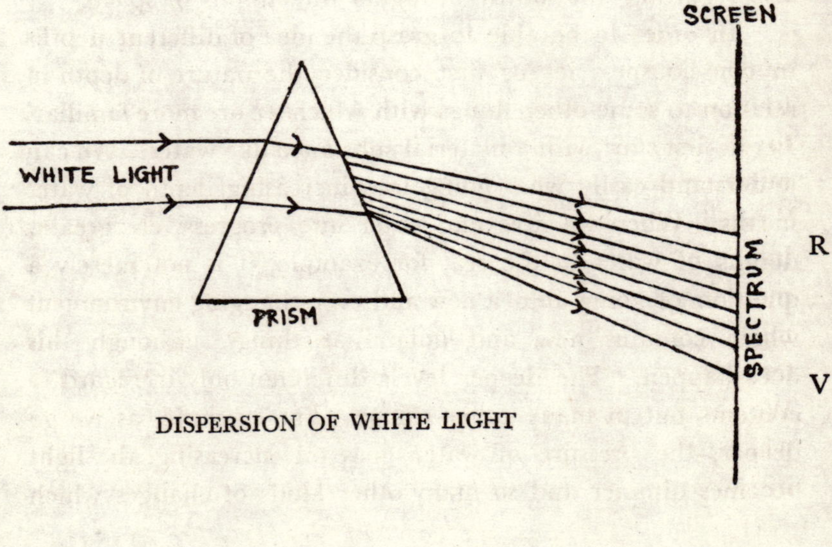

DISPERSION OF WHITE LIGHT

If any one looks towards the prism at the emergent beam from the point R he will see it red because his eye will be in the red region of the spectrum. As he moves his eye downwards he will see the different colours of the spectrum successively until he reaches the point V corresponding to the violet region of the spectrum. If he moves his eye further downwards he will see no colour as the ultra-violet region of the spectrum is invisible.

This moving the eye downwards from the red to the violet end of the spectrum can be considered as moving into greater depths of light because the eye is seeing light of smaller and smaller wavelength, or in other words, light of increasingly subtler nature. White light can thus be considered to exist in different depths and seeing light of different colours as shown above is really movement into greater depths. Another important point to note in the above experiment is that though radiations of smaller and larger wavelengths are present on the two sides of the visible spectrum they are invisible because the eye is not adapted to see them.

The two examples discussed above will have given the student some idea of the significance of depth in relation to things of different categories and we can now proceed to consider the question of different depths in relation to mind or consciousness. For, according to Yogic psychology "raising" one's consciousness in *Samādhi* means really going into greater depths of consciousness. Perhaps it would be more appropriate to say that it is going into greater depths of mind. For, we can go into greater or lesser depth of a differentiated thing only like the spectrum of coloured lights. We cannot really go into greater or lesser depths of white light which is an integrated state and contains all coloured light in a perfectly harmonized state. But as integrated and differentiated states exist side by side in a blended condition, going into different depths of mind can in one sense be considered as going into different depths of consciousness also.

In order to consider the question of depth in relation to mind or consciousness we have to turn to yogic psychology and technique for it is only in the practice of yoga that different levels of mind are encountered and it is possible to transcend all these levels one by one and emerge into the realm of pure integrated consciousness. We need not discuss this question here, for, it has been dealt with thoroughly in other contexts. But there is one point which we may consider briefly on account of its relevance to the question we are considering.

This point which we should note carefully is that if we want to go into the integrated state of consciousness we have to transcend the realm of the mind *as a whole* although this is preceded by advancing from the less to the more subtle states of mind. The process of going from the realm of mind into the realm of consciousness is fundamentally different from the process of going from one level of mind into another subtler level of mind. It is for this reason that the technique of *Nirbīja Samādhi* differs from the technique of *Sabīja Samādhi*. We can again get some help in understanding this difference by considering the above-mentioned experiment of dispersion of white light by means of a prism. If we refer to this experiment we shall see at once that the process of going into different depths of coloured light is very different from the process required for going from the realm of differentiated coloured lights into the realm of integrated white light. In the latter case we have to turn towards the prism which disperses the white light, pass through it and emerge on the other side into the region of white incident light.

In the same way, in *Nirbīja Samādhi* the centre of consciousness has to pass through the mechanism of manifestation, which separates the integrated state of consciousness from the world of manifestation created by the mind, and emerge into this world of Reality. It is this process in its entirety which is hinted at in aphorism IV-22 of the *Yoga-Sūtras* but we need not discuss this question here. The main subject under discussion is the essential nature of dimensions of space.

After considering the essential nature of dimensions of space and their relation to depths of consciousness let us dwell for a while on the mathematical aspect of the problem.

Although higher dimensions had been the subject of study and investigation by mathematicians for a long time, it was only in the beginning of the present century, when the Theory of Relativity was put forward by Einstein that the subject began to be taken seriously. It was necessary for the purpose of this Theory to specify an "event" and this could be done only by specifying the three usual space coordinates X, Y, Z, and a time coordinate, t. The space of this theory is thus a four-dimensional space compounded of three true dimensions of space and one of time. This theory does not, therefore, really deal with the fourth dimension of space as conceived by and known to Occultism. Taken in this general and loose, sense, dimensions of space even higher than four are required in the treatment of problems in other fields. Thus, to specify a moving point, we require a six-dimensional space, three space coordinates X, Y, Z and three coordinates u, v and w, giving its velocity components. In statistics, we need an n-dimensional space, and in quantum mechanics, space of infinite number of dimensions.

It is interesting to note that the mathematical study of higher dimensions was undertaken originally not for its practical applications in the field of relativity, statistical mechanics, quantum mechanics, or other fields of scientific study. The initial motive was pure intellectual curiosity and the thirst for generalization in the mathematician. He wanted to see what would happen if the facts and concepts relating to 1-, 2- and 3-dimensions with which we are familiar were generalized. What kind of worlds would worlds of four and still higher dimensions be if these concepts and facts relating to 1-, 2- and 3-dimensions are extended into the realm of the unknown worlds? How do all these worlds differ from our world? It was only later that these questions studied originally for their purely theoretical interest were found of practical use and the ideas so discovered were applied in many fields of scientific study.

To a mathematician objects of higher dimensions appear as real as objects of 1-, 2- and 3-dimensions though he cannot visualize them. He can discuss their properties and also prove some of them. He can even use his imagination to draw figures of four- and higher-dimensional objects on two-dimensional plane paper in the same manner as a sculptor, architect or a student of mathematics can draw images of solid objects on plain paper.

In order to determine the properties of objects of more than three dimensions and laws from which these can be deduced for different values of n, the usual method is to go to lower dimension (1, 2, 3 . . .) and see how these properties change with the increase in the number of dimensions. In many cases the changes which take place in the properties when we go from 1 to 2 dimensions and then from 2 to 3 dimensions can be generalized and from the laws thus obtained the corresponding properties for any value of n determined or calculated. As the layman will find the mathematical treatment of these properties and laws of objects of more than three dimensions too abstract and difficult we need not enter into this question here. But if any one, with some knowledge of mathematics, goes into this question in detail he will feel convinced that objects in higher dimensional spaces must exist somewhere.

The question of the existence of higher dimensions of space is not merely an academic question. Its chief importance lies in the fact that it alone can provide a reasonable and credible solution of many scientific and philosophical problems. Take, for instance, the difficulty of communicating experiences pertaining to subtler and especially spiritual worlds to those who are still confined to the physical world of three dimensions. Many aspirants feel a grouse that those who have experiences and knowledge of these higher planes of existence do not communicate this knowledge to those who are still in the dark, except in a very sketchy and general form. The sceptic suspects that nothing definite is communicated because there is nothing to communicate and occultists live in a world of make-believe or imagination born of wishful thinking. Neither of these assumptions is right. Nothing much and nothing definite is communicated because of the difficulty inherent in the very nature of these experiences and knowledge. And this difficulty is due mainly to the fact that these higher worlds function in higher number of dimensions of space which we cannot visualize or comprehend.

The organs of cognition and knowledge of each plane are designed to deal with the experiences and knowledge of that particular plane and those who are functioning actively on a particular plane cannot understand or comprehend the experiences and knowledge of the higher planes except in a very superficial manner. In order to gain these experiences and comprehend this knowledge an individual has to unfold the faculties pertaining to these subtler worlds by rigorous self-discipline and yogic training. These faculties are present in a potential form in every human being and can be unfolded by him if he is morally and spiritually mature, is prepared to undergo the prolonged training and make the necessary sacrifices for this purpose.

Then there are many phenomena in the field of philosophy and modern science which cannot be explained in any other manner except on the basis of existence of worlds of higher dimensions. The infinity of expressions following the same pattern, the emergence of forces and objects out of nowhere and without any apparent cause, prevision, and a large number of other phenomena cannot be accounted for or understood except in terms of higher dimensions of space. Sceptics and orthodox scientists may dismiss these things as due to coincidence or accident but anyone who has studied the problems carefully and with an open mind can see that these are merely excuses for maintaining the scientific thesis that there are no other worlds in existence except the physical.

The infinite, the unknown and the unpredictable face us everywhere and most of these things cannot be understood or accounted for except by supposing that subtler worlds of higher dimensions exist hidden within the visible physical world and it is from these worlds that these things which cannot be accounted for emerge either periodically or sporadically. A partial realization of how such things are possible owing to the existence of worlds of higher dimensions can be gained to some extent by considering the relations of worlds of 1- 2- and 3-dimensions which we can comprehend. This method can be used not only for deducing the nature of objects in worlds of higher dimensions by mathematical analysis as pointed out above but also for accounting for the emergence of the infinite, the unknown and the unpredictable into our three-dimensional physical world. Let us consider the relations of worlds of 1-, 2- and 3-dimensions very briefly to gain at least a general idea of how this can be done.

The fact that infinity is closely related to the existence of different dimensions of space and is a matter of perception can be seen to some extent by considering the relation of points, lines and surfaces in worlds of 1-, 2- and 3-dimensions. A line contains an infinite number of points, but as long as consciousness is confined to one dimension it will not be possible to see them simultaneously and realize that they are infinite in number. We can see only one point at a time on either side. How can we see the fact that the number of points in a line is infinite? By transferring our consciousness to a two dimensional world by rising above the one-dimensional line as shown below.

When our consciousness is free to move in a plane containing the line we can see all the points in the line simultaneously and the infinity of points in the line is revealed. But as long as our consciousness is confined to a superficial plane of two dimensions we can see only one line at a time though that line is seen to contain an infinite number of points.

How can we get out of this limitation of being confined to the view of one line only? By transferring our consciousness to a world of three dimensions and being free to move in three dimensions. Then we shall be able to see simultaneously an infinite number of lines in the one surface exposed to our view as shown below:

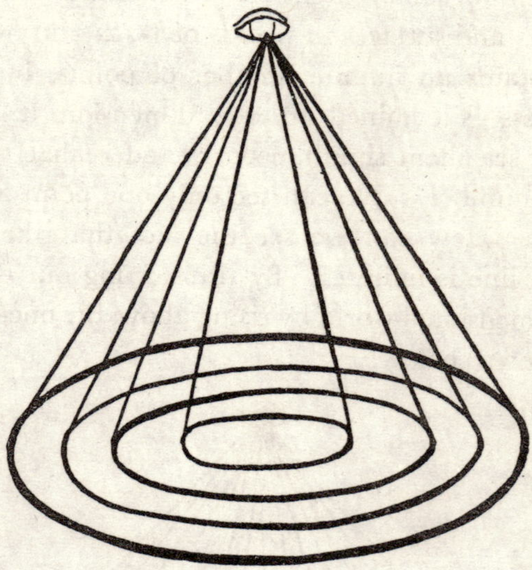

The eye poised above the surface containing the circles can see simultaneously an infinite number of circles or lines in the surface, for, each surface can contain an infinite number of lines. So, by transferring the consciousness to three dimensions the infinity of lines in the surface and the infinity of points in each line is revealed automatically.

It is necessary to note carefully that when our consciousness is functioning in 3-dimensions we can see only *one surface at one time* in this manner. We seem to see many surfaces at the same time but actually these different surfaces of different objects constitute one surface in our consciousness and are seen as one surface, as the student will be able to realize if he considers the problem carefully. That is why ordinary vision lacks the three-dimensional quality. This limitation is partly overcome in a stereoscope or 3-dimensional cinematography by introducing double vision. It is only when our consciousness is transferred to a world of 4-dimensions that we can see a 3-dimensional solid as it is and not superficially, as clairvoyant investigation of the astral plane has shown. On the astral plane we can see a solid object as a solid of three dimensions in the real sense though our consoiousness is functioning in a four dimensional world. It will be seen therefore that we generally see objects in a world of any number of dimensions in one dimension lower than the number of dimensions of the plane itself, if the phenomena of the worlds of 1-, 2- and 3-dimensions are any guide.

We are, however, not considering here these mathematical aspects of dimensions of space in the present context. The important point we are interested in and should note carefully is that at every step in which consciousness is raised into a world of higher dimensions a new world is revealed and this new world reveals new and infinite aspects of the objects which we could see only in a very limited aspect in the world of lower dimension. We also see that these new and infinite aspects already existed, and raising the consciousness to a higher dimension of space merely reveals these new and comparatively infinite aspects. Seeing the different worlds in increasing number of dimensions of space is thus merely a matter of perceiving more and more and newer aspects in a Reality which is eternally whole, infinite and all-inclusive but is perceived partially and incompletely owing to the limitations of the lower planes and our lack of development. The whole of Reality which is referred to as *Pūrṇa* in Sanskrit and exists eternally in its infinite and complete form, is present behind and within every centre of consciousness all the time, and it is only our limitations which account for our seeing it incompletely and in very limited aspects and forms in our limited experiences. Raising consciousness to worlds of higher dimensions merely removes these limitations progressively.

The student should, however, be on his guard against confusing worlds functioning in a higher number of dimensions with worlds functioning on different planes owing to the differences in the size of the particles. In both cases there is lack of cognition of the other world but the cause is different. In raising consciousness into a world of higher dimensions we can see new and deeper aspects in the same things. In the other case, the objective world is different.

The appearance of strange and unaccountable phenomena in a world of lower dimensions on account of happenings which appear quite natural in the corresponding worlds of higher dimensions can also be illustrated and understood to some extent by performing experiments in worlds of 1-, 2- and 3-dimensions. A magnet moving below the surface of a paper on which some iron filings have been strewn will make the iron filings move without any apparent cause to those whose consciousness is confined to the two-dimensional surface of the paper. A bullet fired through the paper will make a part of their world disappear suddenly and without any cause. An inkpot overturned on the paper will make their physical world quite black and moist suddenly and causelessly. Hundreds of such experiments can be devised to illustrate how dimensions of space play tricks on our mind and make us live in extremely limited, illusory and strange conditions of the lower planes. Scientists would do well to ponder these facts and see whether they can throw some light on their puzzling problems and wean them from their philosophy of purely physical materialism.

There is one more aspect of this question of higher dimensions of space which should be recalled before we close the subject and that is the unique nature of zero dimension associated with a point. As has been explained in *Man, God and the Universe* in discussing the nature of the ideal point, an entity whose consciousness is centred in a point with zero dimension is above the limitations of dimension of space and can be aware of all that exists in, and happens in the worlds of different dimensions. The point is thus a common door to worlds of all dimensions. This is, however, only a possibility because in *Samādhi*, although the consciousness passes momentarily through the Common Centre of all vehicles, and thus through the *Mahābindu*, the Yogi does not become omniscient at the moment. Why? Because the real Point is not as simple as it appears to the illusion-bound intellect. It has mysterious aspects of which we can know nothing and there are many kinds of influences which are also acting on consciousness and tending to obscure its vision. The study of these forces and their obscuring influence on consciousness is a part of Yogic psychology and we need not enter into this question here. It is enough to note that the Unmanifest Reality is above the limitations of dimension which characterize " mental spaces " or *Chidākāsha*. In that Reality these mental spaces exist in their integrated form which is referred to as *Mahākāsha*.

The existence of higher dimensions of space and measures of time imparts a new dimension to the concept of relativity and shows how limited and incomplete the modern Theory of

Relativity, which takes into account only three dimensions of space and one of time, is. A total view of relativity will include a consideration of not only worlds of higher number of dimensions but also of that State which is above the realm of time and space. Such a view is, of course, possible only for those whose consciousness can transcend the world of manifestation and can function in the world of Reality.

CHAPTER VII

THE DUALITY OF TIME AND SPACE—II

The Point and Instant as the Basis of Space and Time

As shown in discussing the integrated state of Reality it is a unique state which is indicated by the Sanskrit word *Pūrṇa* which indicates a state which is whole, continuous, complete, the same everywhere. Obviously, there is no place and possibility for the appearance of phenomena in such a state, as it is. Even when an integrated state is differentiated *as a whole*, it still retains partly its integrated character and remains generally a continuous state with no gaps between the different constituents of the differentiated state or within a differentiated state which appears. Thus, when white light is dispersed by a prism, we get a *continuous* spectrum and there are no gaps between the different colours which form part of the spectrum. In other words, the spectrum of coloured lights is a continuous phenomenon. So, mere differentiation is not enough for producing phenomena of all varieties in the integrated state of Consciousness or Reality. What do we require for producing the infinite variety of phenomena which we find in manifestation? Discontinuity, so that things can be separated from one another by means of gaps and the *nirvishesha* condition can thus become *vishesha*.

Since Time and Space form the warp and woof of the fabric which we generally refer to as manifestation we should expect to find discontinuity of a very fundamental character at the basis of these two principles which serve as basic instruments of manifestation. Let us therefore consider carefully the question of discontinuity which lies at the basis of Time and Space both from the Occult and scientific points of view.

According to the Occult doctrine discontinuity is introduced in the realm of Space by bringing into existence an infinite number of points each of which serves as a centre of separation and distinction from the Whole. Since manifestation can take place only through the Great Point the precipitation of an infinite number of such points is possible only by the reproduction or proliferation of the Great Point. From the point of view of Reality this appearance of an infinite number of points by proliferation of the Great Point is illusory, but in the lower world of manifestation these points provide effective instruments for the creation of discontinuous mental images which constitute the basic reality of the manifested world, as we shall see presently.

A point has only position and does not help much in the formation of mental images. But points can be combined into a line by placing a number of them so close together that they give an illusory impression of a continuous line. An illusory impression of a line can also be produced by a rapid movement of a point. Similarly, a number of lines can be placed so close together that they give an illusory impression of a continuous surface. So an infinite number of points or a rapid movement of a point can produce by different combinations an illusory impression of objects in a different number of dimensions. Scientific knowledge fully corroborates this Occult doctrine. For, the apparently solid objects with continuous surfaces are nothing but aggregates of atoms and other sub-atomic particles separated from one another by vast distances. As compared with objects with which we deal in our common experience these atomic particles are comparatively just like points and yet they give us an illusory impression of solid objects with continuous surfaces and no interstices between the particles constituting them.

It should be noted that the illusion of continuity is produced when the interstices between the particles are too small to be perceived by the sense-organs. Each sense-organ is designed and is capable of cognizing impressions within a limited range and fails to register those beyond this range on both sides. It is really a question of relativity —of the relation between the capacity of the sense-organ and the size of the particle or the wavelength of the vibration.

What has been said about space is also true in the case of time. Like space, time is also continuous in the integrated state of Consciousness and thus incapable of producing the phenomena of time. It must be broken up into " instants " or the smallest possible periods or particles of time which are called *kṣaṇas* in Sanskrit. It is these discrete " particles " or ideal " points " of time which underlie all kinds of changes in the spatial impressions and produce the illusory impression of the flow of time in consciousness. These instants or " *kṣaṇas* " like the points of space must be the smallest possible periods or rather " points " of time in their initial stage corresponding and analogous to the ideal points in space. In fact, the Oxford Dictionary defines an " instant " as a " moment " or " point of time ". The importance of breaking up the continuity of time in producing continuous and coherent phenomena can be seen best by projecting a cinematographic picture on the screen without the mechanism which produces periods of darkness intermittently and thus in a way breaks up the continuity of time. Without this intermission which divides time into separate periods a blur instead of a coherent picture will be produced. It is difficult to put these things in words but the student should be able to grasp the underlying idea.

These points of time and space thus lie at the basis of the form side of the manifested universe and together produce all the illusory impressions of the phenomenal world both in its static and dynamic aspects. It is their joint working together in a definite manner and order which underlies and constitutes the basis of the world process in its objective aspect. This underlying process in its mechanical aspect is referred to as *krama* in the *Yoga-Sūtras* and its perception is one of the highest *Siddhis* or accomplishments gained in *Samādhi*. Since this process lies at the very foundation of the mechanism of manifestation its perception is possible only in the last stages of *Samādhi*, when consciousness is transcending this mechanism and passing from the realm of the mind and manifestation into the world of Reality. This process is referred to in aphorisms IV-32, IV-33 and III-53 of the *Yoga-Sūtras*.

What is the significance of the point of space and the instant of time, the fundamental cosmic units of space and time, in the mechanism of manifestation or in the foundation of the universe? It has been pointed out elsewhere that, according to Occultism, space as we experience it on the planes of manifestation, is mental in nature and it is for this reason that it is called *Chidākāsha*. It has also been shown that the mind, whether universal or individual, works through a point which is called *Mahābindu* in the case of a macrocosm and a *Manobindu* in the case of a microcosm. Now, if ordinary space is a mental creation for containing mental objects, a mental phenomenon experienced in the mind, and if in order to fix the ultimate unit of space we try to find the smallest possible amount of space which cannot be reduced and subdivided further, how will we proceed? Let us make a thought experiment as Einstein used to say when he was faced by such a problem. Let us suppose that the individual starts with the boundless mental space which is projected from the centre of his consciousness as an infinite sphere of three dimensions. Let us then suppose that he starts contracting this mental spherical space in his mind gradually so that it assumes the forms of smaller and smaller spheres, all centred in his *Manobindu*, the centre of his consciousness. What is the limit of this progressive contraction to smaller and smaller spheres? Obviously, the ideal point about which they are all centred. If the reduction is sought to be carried further beyond this limit, the consciousness will pass through the centre and emerge on the other side in the next higher plane with its own mental space of a different number of dimensions. If the process is repeated on the successive higher planes one after another the consciousness of the individual will emerge ultimately into the world of Reality in which consciousness is integrated or *nirvishesha* (i.e. without any particular " objects ") and which is the realm of *Mahākāsha* or integrated space.

So, obviously the ideal point is not only the smallest ideal measure of space on the physical plane but on *all* planes of manifestation. And it is not only the common measure of space on all the planes of manifestation but the common centre from which all mental spaces of different planes of manifestation are projected to contain the objects of the respective planes as discussed in *Man, God and the Universe.* And we have seen already how the arrangement of points in one direction produces a line, the arrangement of lines in a certain manner produces a surface and so on. So the point can be and, according to Occultism, is the foundation of the form side of the manifested universe. In fact, theoretically, a single point can produce an impression of a solid object if the movement of the point is sufficiently rapid. Such speeds can be produced even in our laboratories.

How can an infinite number of points needed for producing phenomena in the worlds of manifestation be produced? Just as an infinite number of expressions of an archetype can come into existence when the archetype is projected from the archetypal world into the lower worlds of forms. The archetype of rose, for example, can and has been expressing itself not only in billions or rather in an infinite number of roses, but the roses are *all different from one another* and yet possess the essential nature of a rose. So each point can be not only independent but also unique in its nature and expression. This question is connected with the mystery of the relation existing between an integrated state and the differentiated states which are derived from that integrated state and has been dealt with elsewhere. It is enough to note here that every kind of integrated state can give rise to an infinite number of differentiated states of the same category from the very nature of the relation subsisting between the two states. So the *Mahābindu* can find expression in an infinite number of unique points which can serve as centres of expression for atoms, and individual minds etc., all essentially unique. It is these points which by their aggregation in different permutations and combinations produce the illusory impressions of tangible objects in the worlds of name and form. And it is this uniqueness which is hidden in each atom which expresses itself in and serves as the basis of uniqueness of each mind and each soul.

That all objects which we cognize through our sense-organs can ultimately be a mere aggregate of points in space is possible even on the basis of modern scientific discoveries and theories concerning the nature of matter as has been pointed out already before. For matter, according to Science, is not continuous but discontinuous and its atoms and other particles are separated from one another by tremendous distances. And the conversion of matter into radiation or photons reduces it practically to a number of points associated with energy. The tangible nature of objects and the apparent continuity of their surfaces and volumes is therefore illusory and is due to the fact that the atoms and the interspaces are too small to be perceived individually by the sense-organs of this plane and are cognized by their joint effect *en masse*.

So, we can ourselves see to some extent how the tangible world of objects we cognize through our sense-organs is nothing but an illusory impression created in our mind by the arrangement and interplay of an infinite number of points scattered in the mental space. This shows clearly how the whole of the objective world by which Science swears and on which it has taken its stand is a pure optical illusion or *Maya*. Even scientists can see this patent fact. But what can they do? They dare not go back on their professions and pet theories and the repudiation of the materialistic philosophy which they have been trying to nourish and maintain in spite of all the evidence to the contrary.

Let us now come to the other fundamental factor involved in manifestation—time, which is nothing but the impression produced in consciousness by change of mental images. Is time continuous or discontinuous? As has been pointed out already, according to Occultism it is discontinuous *in manifestation*. It is not only discontinuous but the phenomena of different planes of manifestation are governed by different measures of time related to and based upon a cosmic unit of time which like the cosmic unit of space, point, cannot be subdivided and reduced further. This irreducible unit of time is called *Ksana* or *Kāla-aṇu* in Sanskrit, and like the *Bindu* or Point is an ideal, mysterious entity which eludes grasp by the limited illusion-bound intellect.

What is the relation of this *Kṣaṇa*, the irreducible Cosmic unit of time to the Ultimate Reality underlying the universe? We have seen in the case of the Point that it is the centre of consciousness from which the mental worlds are projected in manifestation. How is the *Kṣaṇa* or the ideal Instant related to this Reality? For an answer to this question we have to go back in thought to the rhythm which underlies the phenomena of the universe and which is rooted in and based upon the Cosmic Rhythm taking place in the Ultimate Reality.

It has been shown in the chapter on Cosmic Rhythm in *Man, God and the Universe* that all natural changes, both in the manifest and the unmanifest, are essentially rhythmic in character and all these different kinds of rhythms are derived from and are partial expressions of the mysterious Cosmic Rhythm taking place at the heart of the universe. Now, the very idea of rhythm involves the idea of a particular measure of time and the existence of definite mathematical relations existing between the different types of rhythms and the measures of time. In Indian music an elaborate technique has been developed for bringing into exact relation the rhythmic movements of sound with the measures of time and this inherent relation of rhythmic movements with definite measures of time can be seen clearly.

If we want to find the smallest possible measure of time in manifestation, which may be called the cosmic unit of time, what will be the most appropriate reality to base our standard upon? Obviously, the fundamental rhythm underlying the universe which has been referred to as Cosmic Rhythm. So the ideal measure of time, the ultimate unit of time —the *Kṣaṇa*—may be taken as the period of time in which a complete cycle of Cosmic Rhythm takes place. This is the duration in which Consciousness expands to infinity from its centre and returns again to the centre after completing a whole cycle, thus embracing the whole realm of existence in its infinite sweep. Let us not go further into the quantitative aspect of the problem which is utterly beyond the realm of the intellect. Let us only try to grasp the general idea behind this concept of *Kṣaṇa* which relates the smallest ideal measure of time with the cycle of Cosmic Rhythm.

This idea is not really as fantastic as it may appear to those who are not familiar with the wonderful developments which have taken place in the related fields of Science. The ordinary unit of time in all scientific work is, of course, the second. This second is based, as is well-known, on the rotation of the earth and for certain scientific purposes is not sufficiently exact and reliable. In their search for a more satisfactory unit of time scientists have discovered that the vibrations of atoms provide a more reliable and exact standard of time than the comparatively crude standard provided by the rotation of the earth. This is natural because these vibrations must be related to Cosmic Rhythm and must be constant. The vibration period of Cobalt is now generally used for fixing the standard of time though this is fantastically small and requires an elaborate technique for its measurement. Such devices for measuring time in all scientific work of the highest order are called Atomic Clocks.

If we have grasped the principle underlying the modern devices for measuring time by vibration periods we shall be able to see not only the significance of *Kṣaṇa* but also appreciate the wonderful conception of time which is contained in the Occult doctrine of *Kṣaṇa*. Upon this mysterious unit are based the measures of time of the different planes of manifestation, these measures depending upon the plane and being related not only to one another but also to the dimensions of space governing the corresponding planes.

How is this ultimate unit of time—the *Kṣaṇa*—related to the measures of time operating on the different planes of the universe? It is not possible for us to understand this relation while we are still in the realm of time. As pointed out in the *Yoga-Sūtras* (IV-33) the whole process of manifestation involving time becomes perceptible only when consciousness is released from the limitations and illusion imposed upon it by time. And as time, in its different measures, pervades the whole five-fold field of manifestation, the process can be perceived and really understood only when consciousness is emerging from the world of manifestation into the world of Reality. In fact, so closely and intimately is the World Process related with the process of time based on the succession of *Kṣaṇa* that *Saṃyama* or contemplation on the time process involving *Kṣaṇas* releases consciousness from the World Process and enables it to emerge into the World of Reality which is above time and space. This point has been made clear in aphorism III-53 of the *Yoga-Sūtras*.

But though we cannot perceive or even understand this process with our present limitations it is possible to gain a general idea of the principle underlying the existence of different time measures regulating the phenomena of the different planes of manifestation. All these are derived from *Mahākāla* the principle of Time existing in the Ever-Unmanifest by a process of differentiation and all are based upon the ultimate unit of time, the *Kṣaṇa*. This question has been discussed in some detail in Chapter XXIX of *Man, God and the Universe*.

It is this Cosmic Rhythm and the measures of time which are based upon it which are symbolized by the *Damaru* of *Mahesha* in Hindu symbolism. The *Damaru* symbolizes not only the rhythmic nature of the movement which underlies the dynamic aspect of manifestation but also the intermittent nature of time which is derived from it. It also symbolizes the vibratory nature of the Energy which is produced by this rhythmic movement and which is called *Nāda* or " sound ". A careful study of this symbol will not only show us how effective it is in representing some of the important aspects of the mechanism which underlies the universe but will also help us to understand to some extent this mysterious mechanism.

The intermittent nature of time is a philosophical and scientific concept of the greatest significance and can be illustrated almost perfectly by the projection of a cinematographic picture on a screen. Although the projection produces the impression of a continuous series of events, we know that this continuity is merely an illusion produced by our inability to distinguish between the alternate periods of illumination and darkness. The manifested universe is similarly an intermittent phenomenon owing to the intermittent nature of time which brings about changes in it. But instants or moments of time succeed one another so rapidly that we cannot distinguish between the periods in which the universe appears and disappears alternately. The procession of events which we cognize through our sense-organs or instruments of cognition are thus seen as a continuous phenomenon but this continuity is illusory.

The appearance and disappearance of the universe may be considered as the result of the turning inwards and outwards of the Consciousness of *Mahesha*, the Presiding Deity of the manifested system. When His Consciousness is turned outwards, the manifested system comes into existence. When His consciousness is turned inwards towards the unmanifest the manifested universe disappears. It is this turning inward and outward of His Consciousness which not only enables Him to bring together the unmanifest and the manifest but also fixes the measure of time for the particular universe over which He presides. A complete cycle of this inward and outward turning of consciousness may be considered as a *Kṣaṇa*, the fundamental unit of time for the system.

To this intermittency of time may also be attributed the fundamental fact discovered by physicists that radiant energy is emitted or transferred not in an unbroken stream but in discontinuous bits or portions which are termed *quanta*. Each quantum carries an amount of energy given by the equation $E=nv$ where v is the frequency of the radiation and n is Planck's Constant. If energy comes from its infinite source in the unmanifest and time is intermittent it is inevitable that the energy should break up into separate and discontinuous portions or quanta. If water is being pumped from a reservoir by means of a water pump and the pump works intermittently the water supply will become divided into separate portions. So the Quantum Theory not only throws light on the nature of the mechanism which underlies the manifested universe and which is referred to as *krama* but partially corroborates this Occult doctrine.

We have seen that the Principles of Consciousness and Power which are referred to as *Shiva* and *Shakti* in Hindu philosophy are the basic Realities of all manifested worlds and the universe may be considered as the expression of this dual Principle. If this is so then every duality of a fundamental character in manifestation must be rooted in and should show the essential character of the primary duality and this is really the case even though we may not be able to trace the underlying relation owing to our limitations.

We shall see later in the case of the duality of matter and energy how it reflects and expresses in a remarkable manner the primary duality of the *Shiva-Shakti Tattva*, matter reflecting and expressing the static principle of Consciousness and energy the dynamic principle of Power. Since Time and Space constitute another fundamental duality, perhaps the most fundamental duality in manifestation, we should expect it to show the same kind of relation to the primary duality of Consciousness and Power. In other words they should show the intimate and indissoluble relationship of a polar duality and should reflect and express the two constituents of the primary duality.

If we consider this question carefully and examine the nature of Time and Space in the light of the Occult knowledge concerning the nature of *Shiva-Shakti Tattva* there will be no difficulty in seeing that *Shiva* is related to Space and *Shakti* to Time. *Shiva* is the static principle and *Shakti* the dynamic principle in the dual *Shiva-Shakti Tattva*. The former is potential, changeless, stable while the latter is kinetic and the motive power of movement and change.

Leaving aside the Occult conception of Space and Time for the present, even if we consider these two realities underlying the universe in the light of our ordinary experience we shall see easily that space as we conceive it has all the attributes associated with *Shiva*, the principle of Consciousness. It is potential and capable of containing all objects which can possibly be imagined by the mind, or according to our common notions, can exist on the physical plane. It is changeless because while the objects change, the space which contains them remains the same. It is stable because it pervades and transcends all objects and is not at all affected by them or by the changes which take place in them.

On the other hand, Time has all the attributes which are associated with *Shakti*, the principle of Power. It is the dynamic element of the Space-Time duality. Wherever there is movement time is always associated with it. In fact, movement is always measured in terms of the time unit in all scientific work. Wherever there is change time comes in inevitably, so much so, that there is no way of perceiving or measuring time except through changes in the objects in our environment. Even Science now recognizes this fact and it is one of the basic ideas upon which the Theory of Relativity is based.

Not only are Space and Time related to *Shiva* and *Shakti*, not only are they bound together by a polar relationship like *Shiva* and *Shakti*, but even their products in manifestation like matter and energy are related to *Shiva* and *Shakti* and have thus the fundamental characteristics associated with Consciousness and Power, as has been pointed out already. We can see this easily if we consider the basic realities of the physical world. What are these basic realities? Matter and energy, or in scientific terminology, mass and radiation. What is matter according to the Occult doctrine? It is a condensed form of *Ākāsha* associated with energy. The derivation of all states of matter from *Ākāsha* is shown as follows:

Ākāsha⟶*Tejas*⟶*Vāyu*⟶*Jala*⟶*Prithivi*
(*Vāyu*) (*Tejas*)

The gaseous, liquid and solid particles provide the tangible and multifarious vehicles for matter. They partly determine the properties of objects composed of them. The energy associated with these particles in the form of atomic energy or molecular motion provides the dynamism of these particles upon which their properties and interaction depends. The atomic energy locked up within the atom determines the properties of the atom. The molecular energy associated with the atoms and molecules determines their interaction and effect upon the sense-organs. The energy of radiation which itself is without any material basis or association can, in the form of photons and light waves, also affect the material particles and the sense-organs. How material particles are the expressions of the *Shiva Tattva* and radiation of the *Shakti Tattva* has been shown in another context.

CHAPTER VIII

THE DUALITY OF MATTER AND ENERGY—I

Materialism without Matter

It is the duty of all serious students of Occultism not only to disseminate the truths of the Divine Wisdom in the world as widely as possible but also to influence and guide as much as possible the development of thought in those fields which are of vital importance to humanity. These truths of the Divine Wisdom are not meant only for those who are prepared to accept them and utilize them for their inner unfoldment. They are meant for humanity as a whole. This is quite clear from the fact that the Great Ones who are the guardians of humanity attach as much, if not more importance to the work for the evolution of the humanity as a whole than to the spiritual unfoldment of individuals who are sufficiently mature for this purpose.

The phenomenal development which is taking place in different fields of Science is a proof of the keen interest taken by the Great Ones in this development and the importance which they attach to this work. There can be no doubt in the minds of those who have been watching this rapid and unexpected development that it is being guided from above. The manner in which the extraordinary discoveries made in different fields of Science during the last fifty years have been prepared for and coordinated systematically shows the guiding hand and intelligence which has made these discoveries possible.

It may appear rather strange to those who believe in the spiritual basis of the universe and the incomparable superiority of the spiritual outlook over the prevailing materialistic outlook, that while such extraordinary and effective help is being given from above for this scientific development, the help for the moral and spiritual evolution of man is comparatively poor, and moral and spiritual evolution seems to be lagging far behind intellectual evolution.

When we feel inclined to judge the wisdom of Divine guidance by our standards of ignorance it is good to remind ourselves that our vision is too limited and our knowledge is too meagre to enable us to form a correct idea of the significance of the prevailing trends and the nature of the future developments. We see frequently in our individual life that an incident which appears like a disaster in the present proves to be a blessing in the future. Similarly, many developments in the world which at the time appeared undesirable and even dangerous prove ultimately of far-reaching benefit to the human race.

The extraordinary and spectacular development in many fields of Science which has taken place during the course of the present century is a case in point. It has placed tremendous power in the hands of man who is still morally and spiritually undeveloped. It has led to an extraordinary growth of the materialistic outlook and preoccupation with the material needs of man. Looked at from our narrow point of view, these developments may appear undesirable and even dangerous. But who can say what form they will take in the future? The control and utilization of the forces of Nature may be freed gradually from its evil potentialities and usher in an era of unprecedented prosperity and contentment. This will provide the requisite basis of social stability which is needed for the moral and spiritual evolution of man.

The growth of the materialistic outlook and the abundance of physical amenities and luxuries provided by Science may give humanity the opportunity to drink the cup of physical satisfactions and pleasures to the last drop and realize that these satisfactions are illusory and ephemeral in their nature and cannot provide real and lasting satisfaction to the real man imprisoned in the physical body. This surfeit and consequent dissatisfaction with a life of purely physical satisfactions will inevitably lead at least the more intelligent section of humanity to search for happiness in new directions and incline them to give a trial to the spiritual ideals which they have ignored. Signs of this reaction against a life of mere physical comforts and enjoyments, which is essentially a life of the animal, have already begun to appear and these may gradually grow into a pronounced swing in the direction of the spiritual outlook and a life guided by spiritual ideals. Nature is a hard task-master and if we cannot learn a needed lesson through wise discrimination makes us learn it through bitter experience.

An unexpected development of the greatest significance has already resulted from the extraordinary and spectacular growth of Scientific knowledge. In its search for a Godless reality which underlies the unreal world of sense-perception Science has found merely an elusive world of mathematical symbols and formulae which are utterly incomprehensible and meaningless. The world of matter and energy on which it had erected its edifice of scientific materialism has disappeared into a phantom world of mathematical abstractions.

Having rejected in the very beginning the idea of the independent existence of mind, Consciousness and a Divine Reality underlying the material universe, Scientists are now feeling confounded and confused by these unexpected developments but are not yet ready to do some rethinking with regard to their basic assumptions upon which scientific materialism is based. They are taking refuge in all kinds of strange ideas which can be seen easily to be nothing but excuses to maintain their orthodox points of view. They are trying to compensate for the inadequacies of their philosophy by an abnormal and almost pathological concentration on the development of technology. But how long can they resist the logic of facts which they themselves have discovered and the necessity of a more reasonable and satisfactory philosophy for making human life purposeful and meaningful. The human Spirit cannot be kept satisfied forever with an assortment of scientific gadgets and mathematical abstractions which nobody can understand, not even those who have discovered them.

In the present mood of scientists and the chaotic conditions prevailing in the world at present it is perhaps not possible to do more than state clearly the implications and significance of recent scientific discoveries and relate them as far as possible with the Occult doctrines. The powers that have brought about the favourable developments in Science will also, in due time, provide the necessary channels of communication between Science and Occultism. But before we can utilize these channels which may be created, we should be clear in our own mind about the significance of these scientific discoveries and their bearings on Occult doctrines.

Since most of the important scientific developments which have taken place in the present century have found their expression and integration in Einstein's Theory of Relativity this theory and its conclusions have a special significance for the student of Occultism. In this theory he can find many ideas which partially corroborate and throw light on many important Occult doctrines and thus make the study of these doctrines more thorough and meaningful. He can also discover in this Theory many points which can serve as bridgeheads for building bridges between modern Science and ancient Occultism. Scientists may not feel inclined to use these bridges in their present state of mind but if the bridges are there it will be easier to establish communication in due course of time. For, Science will have to come to Occultism ultimately for the solution of its insoluble problems and take to the Occult method for finding the Divine Reality which is hidden within the heart of the universe. The search for the reality of the materialistic conception is doomed to failure for it is like the proverbial search in a dark room for a black cat which is not there.

So, let us briefly review in outline some recent developments in Science which are of special interest to the student of Occultism. It is not necessary for our purpose to go into the mathematical and technical aspects of these developments which the ordinary student cannot understand without acquiring the basic knowledge of Science and higher mathematics.

Since the fundamental point of contention between Occultism and modern Science is with regard to the ultimate basis of the universe—the former maintaining that it is spiritual and the latter that it is material—let us first review the scientific developments which have led to the conclusion that matter, as Science conceived it originally, does not exist and so scientific materialism has lost its material basis. According to present knowledge the ultimate and only basis of the universe is radiation either "bottled-up radiation" which we call matter or free radiation which is known as light or other electro-magnetic waves of different frequencies. There have been many other interesting developments along different lines but this is the most important because it has deprived scientific philosophy, which believes in concrete things only, of its very substance and left it floundering in a morass of abstract ideas. As this is a very important development from the Occult point of view in its contention with scientific materialism it will be worth while reviewing the sequence of scientific discoveries which have led to this unexpected and startling conclusion.

Naturally, since the whole fabric of the universe has been reduced to radiation the problem started with the study of radiation. Everybody knows that when heated bodies become incandescent they emit a red glow that turns to orange, then to yellow and finally to white as the temperature increases. All efforts to find how the amount of energy given off by such heated bodies varied with wavelength and temperature failed until Planck found by pure mathematical analysis his famous equation $E=nv$ which satisfied the results of experiments. He made the unwarranted and strange assumption that the radiant energy is emitted not in an unbroken stream but in discontinuous portions which he termed *quanta*. He had not the slightest evidence for such an assumption but made the assumption because it was only on this basis that his empirical equation $E=nv$ could be made to tally with the results of experiments. It was one of those cases in which assumptions are chosen to suit the conclusions which have been arrived at on the basis of experimental results. The trick lies in working backwards from the conclusion to the original assumptions and choosing those particular assumptions, however strange or unjustified they might be, which will lead to the predetermined conclusion. This is not unlike putting the cart before the horse in the reasoning process and the device is adopted to give to the theory an apparently reasonable appearance when actually it has no basis in reason.

Herein lies the weakness of the scientific theories and the philosophy of materialism which is based upon them. Scientific knowledge is essentially empirical and the veneer of reason which is generally provided is to give it a respectable appearance in the eyes of the public. Scientists are concerned mainly with the "how" of the natural processes and have always and persistently avoided facing their "why" aspect. That is the source both of their strength and weakness. Because they have a reliable knowledge and complete control of the "how" of the processes they can manipulate these processes perfectly and utilize them in producing extraordinary and spectacular results. And because they have not seriously tackled the "why" of these processes and are not really interested in it, owing to the absence of the true philosophical spirit, they have not been able to gain true knowledge with regard to the real nature of anything and the whole field of scientific thought is strewn with the wrecks and remnants of ever-changing and out-moded theories. But the scientists have a firm grasp over their empirical facts and by continuing to produce spectacular results can always camouflage the weaknesses and the inadequacies of their philosophical thought.

This discovery of Planck's Constant by this irrational and apparently unjustified method also incidentally throws some light on the manner in which scientific thought, or as a matter of fact, any kind of development in human affairs is guided from above or from behind the scenes. An idea or some hint with regard to a course of action is put into the mind of the person who is engaged in the work, is adopted by the person concerned without any rhyme or reason, sometimes without any justification, and the development of thought or movement is assured sometimes with far-reaching consequences. The descent of such ideas and solutions from above is recognized even in scientific circles, but they dismiss such occurrences as cases of " intuition ". Nobody asks what intuition is and how such important ideas, sometimes apparently irrational and unjustified are received from nowhere and adopted by hard-headed intellectuals without question. Planck's Constant has dominated the computations of atomic physics for half a century, but it owes its discovery not to the cold logic of reason but the mysterious fluke of that much suspected faculty called " intuition ".

On account of the lack of justification and apparent absurdity of Planck's fundamental assumption—that radiant energy is emitted in discontinuous bits or portions—very few scientists took Planck's Quantum Theory seriously till 1905. In that year Einstein published his Theory of Relativity to account particularly for the puzzling phenomenon which was referred to as the " photo-electric effect ". Einstein postulated in his Theory that all forms of radiant energy—light, X-rays, etc., travel through space in discontinuous quanta and so, practically adopted and incorporated Planck's Quantum Theory in his Theory of Relativity. It was this fact which showed the far-reaching implications of Planck's Theory which was till then regarded merely as a clever conjecture. According to Einstein all light or rather radiant energy is composed of individual particles or grains of energy (which he called photons) and the greater the frequency of the radiation the more energy is contained in its photons. If the wavelength could be small enough, one single quantum could have all the energy of the universe, theoretically.

Einstein's theory that light consists of discontinuous particles was in direct conflict with the then prevalent theory that light is made up of waves. The theory that light consists of electro-magnetic waves was based firmly on the existence of the phenomena of diffraction and interference which are strictly wave characteristics and could not possibly be shown by particles of any kind. And yet Einstein's Photo-Electric Law showed that light must consist of photons or light particles. Which of these two diametrically opposite views represents the actual facts? This fundamental question—whether light consists of waves or particles of energy—has not yet been answered in spite of a tremendous amount of research work having been done by topmost scientists. In fact, the deeper they try to go into the problem the more baffling it becomes.

It will be seen from the above how these unexpected and startling developments had the effect of abolishing completely the distinction between matter and radiation (or energy).

It is interesting to note how physicists have tried to resolve the paradox presented by waves of matter and particles of radiation. They now maintain that it is not necessary to consider the properties of an individual electron or any other fundamental particle of matter. For, in experimenting with these particles in the laboratory we do not deal with single particles at all, but with billions of particles *en masse*. As we have to deal with mass behaviour the problem can be tackled by the statistical method which is based upon the laws of probability. Two German physicists, Born and Heisenberg, developed mathematical expressions which enabled the physicists to interpret quantum phenomena both as waves or particles. But in this ingenious method to resolve the paradox the incomprehensible "waves of matter" were made still more incomprehensible "waves of probability".

And so the modern physicist thinks that he has resolved the paradox of matter behaving like waves and radiation behaving like matter. But to the intelligent man who thinks that no problem can be considered as solved unless the solution is at least intelligible, the above developments seem to have made the nature of matter and radiation and their mutual relationship far more mysterious. To provide a set of mathematical formulae for a certain phenomenon is no more solving the problem than writing $\sqrt{-1}$ is resolving the mystery of the square root of -1. As everyone knows, mathematics defines only the relations between things and leaves the question of their nature untouched. Mathematical formulae can therefore only define and help us to understand the *processes* of nature. They can tell us how things act and behave and not what they are. As Science is interested only in the "how" of things, so that it can exploit them, these mathematical formulae may be adequate for its purpose but to claim that they resolve the paradoxes posed by Nature, as is frequently done, means that we do not really know what "knowing" means.

To "know" things as they are we cannot evade the question of their "why" as is done by scientists. It is true that when we proceed to know the "why" of things the answer to every question which we solve raises another question in its place and so the search for a solution seems to be interminable. But for the serious seeker this is really not so. For there is a Reality hidden within the universe and this Reality contains within itself a real solution for every problem and a counterpart for every illusory and external fact of existence. To " know " things as they really are and to resolve paradoxes posed by them we have to penetrate through our mind and consciousness into this world of Reality. Then we can " know " in the real sense the partial reality corresponding to any definite fact of existence or the Whole Reality in which all these partial realities are contained in a perfectly harmonious relation. The technique of this discovery as is well-known to the students of Yogic psychology is called *Sabīja* and *Nirbīja Samādhi.* But it is too much to expect hard-headed orthodox scientists to adopt or even to consider these unorthodox methods of " knowing " the Truth when all they are really interested in is the phenomenal side of Nature and the exploitation of natural processes for their extremely limited purposes.

The Reality which is the basis and source of the physical universe is of the nature of consciousness and so it cannot be discovered or known by physical methods or by mathematical analysis. It can be known only by our consciousness, or in our consciousness, by methods which are appropriate to the uncovering of consciousness. So, the hope of the scientific world to find the underlying reality of the universe by physical methods aided by mathematics is doomed to remain unfulfilled and they will have to remain contented with the meaningless and incomprehensible " reality " of mathematical abstractions and equations.

How long man can remain satisfied with these abstractions and continue to regard himself " merely as an ephemeral conformation of the space-time field " is for the future to show. But if the Divine Life is present within the heart of every human being, and he is destined to know his Divine nature, it will not allow him to remain in this state of confusion for ever. A time must come when he throws off the shackles of scientific materialism, asserts his Divinity and starts to find his Real nature in the only way this can be done.

CHAPTER IX

THE DUALITY OF MATTER AND ENERGY—II

Energy as the Basis of the Universe

We have seen in the last chapter that matter and energy constitute a duality according to Science, being interconvertible and constituting one reality underlying the physical world. Science stops at this idea, and though it has investigated the behaviour of matter and energy very thoroughly and can utilise them in innumerable ways with perfect confidence, it does not go very far in understanding the essential nature of these two basic constituents of the physical world. Occultism, on the other hand, has not investigated with such precision and thoroughness the behaviour of physical matter and energy, but it has investigated and knows with certainty their essential nature, not only on the physical plane but on all the planes of the manifested universe. But before we deal with this occult conception let us first summarize some simple relevant scientific facts with regard to matter and energy in their physical aspect.

1. According to Science energy is of two kinds, potential and kinetic. Kinetic energy is either vibrational as in the case of radiation or molecular as in the case of motion of atoms and molecules. How are the two related? They are interchangeable. Vibrational energy of radiation can

set molecules and atoms in motion as happens when sunlight falls on physical objects and they get heated. Conversely, metals and other objects when heated begin to give out radiant heat and when heated beyond a certain limit also light. So, radiant and molecular energy are interchangeable.

2. Potential energy, according to Science, is energy in which no movement is involved and it is based upon *force* as kinetic energy is based upon *motion* or movement (vibrational or molecular). What is the nature of the force? It may be force of gravitation or any other force like that present in a coiled spring. In the atoms and molecules, various kinds of forces of attraction and repulsion etc. are locked up in equilibrium and provide the potential energy, while the movements of electrons and atoms provide the kinetic energy.

3. In considering the duality of matter and energy we should note that energy is the basic element and matter is derived from the condensation or congealing of energy. In one sense the two elements of a duality may be considered on a par since both are present in manifestation in a state of equilibrium but looking at them from a deeper point of view we can see that one of them is the basic or essential constituent. Thus taking the duality of time and space, we can see that space is the basic element of the duality, for, there must be mental images in *chidākāsha* or " mental space " before those mental images can change and produce the impression of time. Similarly, we can see that in the duality of matter and energy, energy is the basic constituent. For, we can imagine all matter to be converted into energy and remain in that form without any admixture of matter but we cannot imagine matter without energy, for, if all energy is converted into matter, energy will still remain in matter in a congealed form.

Let us now consider what Occultism has to say about the duality of matter and energy. As far as the relation of matter and energy is concerned the occult conception is broadly the same as that of Science. Both regard matter as a derivative of energy produced by the condensation or "bottling-up" of energy. But Science knows practically nothing about the source of this energy. It cannot explain where the tremendous amount of energy found in the universe, both in the form of matter and free energy, comes from and how it was raised to the high potential which it must have had when the universe began its career. There is no dearth of theories with regard to the nature and origin of the physical universe, but these are speculations based upon very meagre and uncertain physical data. The discovery of quasars and of objects in distant galaxies travelling with speed greater than that of light has resulted in utter confusion of thought in the scientific world and cast serious doubts on the validity of modern theories with regard to matter, energy, and the nature of the universe.

The Occult Doctrine with regard to these fundamental questions is quite clear. According to it the whole of the manifested universe, in its objective aspect, is a product of Energy which is an expression of Divine Power. It is this Energy which in its infinite permutations and combinations gives the illusory impression of all kinds of forms and of an objective world in the mind. It is this Energy which, on losing its dynamism progressively, assumes increasingly dense states of so-called matter.

How this Energy is derived from Divine Power is not easy to understand because the whole complicated mechanism of manifestation is involved in this question and this mechanism can be known in the real sense only when we are able to transcend it through the practice of Yoga as pointed out in aphorisms III-53 and IV-33 of the *Yoga-Sūtras*. But let us dwell for a while on this interesting question so that we can at least gain a general idea of the principles which are involved in the appearance of an objective world in consciousness.

It has been pointed out in other contexts that power, the potential source of energy, is an attribute of consciousness, inseparable from it and related to it in a mysterious manner as a polar opposite. Every manifestation of power or energy must therefore be associated with, and have its source in, consciousness. And as consciousness can express itself only through a mathematical point all manifestations of power or energy must, at their source, be associated with mathematical points and find expression through such points. Every expression of energy must be associated with its own specific point and cannot exist apart from such a point just as every limited expression of consciousness must have its own specific centre from which it is projected in the realm of manifestation.

Recalling what has been said in a previous chapter with regard to the proliferation of centres of consciousness from the *Mahābidu* and the gradual unfoldment of consciousness through them, and also the fact that consciousness and power are indissolubly linked together, we can see easily that even in the initial stage of the appearance of an individual centre from the *Mahābindu* these two aspects of Reality, Consciousness and Power, must be present together side by side. On the one hand, each such centre must be initially a centre of consciousness with infinite potentiality for unfoldment of mind and on the other hand, it must be a centre of power through which can find expression an infinite variety and amount of energy. And both these potentialities must, by their very nature and relationship, unfold *pari passu*.

It is true that it is power which is the correlate of consciousness and not energy but when we are dealing with the expressions of integrated consciousness as the differentiated states of mind we have also to deal at the same time with the expressions of integrated power as the differentiated forms of energy. In the realm of manifestation we have to deal mostly with the expressions of mind and energy though these are rooted in and derived from integrated consciousness and integrated power respectively.

It follows from what has been said above that each ultimate unit of such expression, such as an electron, photon or atom must be simultaneously a rudimentary expression of mind and energy. Just as all expressions of mind are centred round the " I " and co-ordinated by the individualized centre of consciousness so all expressions of energy are centred round the centre of mind and cohere in that centre and can find expression wherever that centre of mind manifests. When an individual incarnates, it is not only his mind which finds expression in his life according to its evolution but all the energies—mental, emotional and spiritual—which have been released during the course of his evolution. In the case of elementary particles like the atoms or their combinations in the form of various kind of organisms we cannot see the expressions of mind or consciousness but only that of energy. Manifestation thus appears to be a grand display of Divine Power or *Shakti* expressing itself in myriad forms of energy. Divine consciousness and mind, being subjective, remain in the background, invisible. It is only those who can dive into the deepest layers of consciousness where these energies have their source who can perceive the Divine Mind and Consciousness behind these expessions of Energy.

The Divine Power which lies at the basis of the manifested universe expresses itself in innumerable ways corresponding to the countless functions of Divine Consciousness in manifestation, such as the powers of love, knowledge, creation, destruction, illusion etc. The powers of producing motion or force from which kinetic and potential energies are derived are only two of these countless powers. But these two powers, finding expression in innumerable permutations and combinations of energy, provide the basis of the objective universe as has been explained in other contexts.

This Divine Power which is inherent in Divine Consciousness and is the source of all kinds of energies needed in a manifested system is infinite, both qualitatively and quantitatively, because it is an integrated state. When we use the word infinite in relation to such a state we mean really "infinite" and not merely tremendous or very great. That is how *any* amount of energy at *any* potential which is needed in a manifested universe can come forth from the integrated Divine Power which is present potentially in the Unmanifest. This fact is pointed out very clearly in many aphorisms found scattered throughout occult literature such as the following two:

Ananta shakti khachitam Brahma sarveshvareshvaram

"Brahman, the One Reality, who is the Supreme Ruler even of all the *Īshvaras* in a manifested universe is endowed with infinite and boundless Power"

Tadā Sarvāvaraṇa—malāpetasya jñānasyānantyāj jñeyam alpam.

"Then, in consequence of the removal of all obscurations and impurities, that which can be known (through the mind) is but little in comparison with the infinite knowledge (obtained in Enlightenment)" (Y.S. IV-31)

Consciousness and Power being correlatives or polar opposites, integrated consciousness and integrated power must exist at the same level and should go together. Similarly, knowledge in the realm of the mind and powers in the realm of manifestation, being differentiated products of integrated consciousness and power respectively, should also be correlatives.

It is a well-known doctrine of Occultism that manifestation is a Self-imposed limitation on Reality. It follows from this that when infinite integrated power descends into the realm of manifestation it loses its infinite character and becomes finite however great it may be. Similarly, when the infinite knowledge present in the world of Reality descends into the realm of the mind it loses its infinite character and becomes limited however great it may be. The two Sanskrit quotations, given above, bring out clearly the distinction between the two states of knowledge and power existing in the realm of the manifest and the unmanifest.

The first aphorism points out that the Power of Brahman, the transcendental Reality in the Unmanifest, is infinite and, as knowledge is a correlative of power, the knowledge present in that state of consciousness must also by implication be infinite. In fact, this power is that of Brahman and infinite power can be associated only with infinite knowledge.

Since all the *Īshvaras* function in the realm of manifestation and are omniscient and omnipotent, as far as their particular system is concerned, their knowledge and power though tremendous must be relatively finite. This distinction of both—knowledge and power—in the realms of the Unmanifest and the manifest is indicated very clearly in the *Yoga-Sūtras* of Patañjali.

It is also a well-known Occult doctrine that the Logos of any manifested system derives both his plan and the power to execute that plan from the Unmanifest. The source of such a universe which is boundless and contains countless systems in manifestation must necessarily have infinite knowledge and infinite power. Not only this, but that Brahman the Unmanifest Reality must be the Supreme authority exercising in a mysterious and invisible manner sway over all the manifest Deities.

The second aphorism from the *Yoga-Sūtras* given above points out the same truth from the point of view of consciousness. It emphasizes the important fact that the nature of knowledge in the two states—manifest and Unmanifest—is fundamentally different. The one referred to as *Jñeyam* is *alpam* or limited while the other is *ananta* or unlimited or infinite. It is not only infinite but of a different quality altogether. The integrated state of knowledge not only contains all the differentiated states in a potential form but is of different character, just as white light is remarkably different from all coloured lights and is not merely their summation.

Those who have made a study of the *Yoga-Sūtras* must have noticed that Patañjali deals with all the questions of spiritual life from the point of view of consciousness, knowledge or perception and the presence of power is only implied in all all that is said about knowledge. It is only in dealing with omniscience that omnipotence is also definitely mentioned as a correlative (aphorism III-50). The fact is that the path of Yoga is essentially a means of gaining knowledge concerning the superphysical states of existence and the Yogic powers are considered only as by-products of this knowledge and not aimed at directly. This truth is clearly emphasized by the fact that the Yogi has to renounce all powers, and in the last stage even omnipotence, before he can gain Self-realization and enter the world of Reality.

On the other hand, in the *Tantras* Reality is generally seen and sought as Divine Power, and the knowledge which is associated with that Power and is attained by becoming one with that Power, is only implied. It is true that in some of the *Tantras* there is an unhealthy emphasis on the acquisition of different kinds of powers, sometimes by questionable means, which has made them suspect in the eyes of the students of occultism. But the highest philosophy and technique of *Tantras* is really concerned only with the question of Self-realization and though this is sought to be attained by the invocation and intervention of Divine Power, it is knowledge of Reality which is the object of *Sādhana*.

How the Ultimate Reality which is pure consciousness in the subjective state assumes, step by step, the state of energy in the form of forces and motions of different kinds and thus provides the objective basis of the universe, is a question which involves the whole field of the Unmanifest and the manifest, and is too transcendental and complicated for the limited human intellect to grasp. It is a question which is resolved in consciousness when the individual is able to free himself from the veils of the mind and becomes established in the world of Reality. This question involves really the mechanism of the world process in its entirety and can thus be resolved only when the world process is transcended in the last stage of Self-realization. This fact has been clearly indicated in aphorisms III-53, IV-32, IV-33 of the *Yoga-Sūtras*.

But though we cannot grasp the nature of this world process in the real sense a glimmer of the truth underlying it can be obtained from a deep study of the Occult doctrine concerning the One Reality and its descent, step by step, into lower and lower states of itself. This question has been discussed in other contexts and it has been shown that the main links in this process of descent may be broadly indicated as follows:

Reality —> Consciousness —> Mind —> Energy —> Matter

From this Energy is elaborated the five-fold universe through the agency of the *Panch-bhūtas* generally referred to as the Five Cosmic Elements in English.

What has Science to say about the role of energy in the universe? Recent developments in the field of physics and chemistry have shown that matter and energy are not two separate realities and that matter is nothing but " bottled-up " energy. It follows from this that the essential basis of the universe is nothing but energy and the objective universe is thus an energy phenomenon. As far as this principle is concerned Science and Occultism are therefore in broad agreement.

But the researches in quantum mechanics have brought Science still closer to Occultism and have further corroborated the Occult doctrine with regard to the role of energy in manifestation. It is not possible for the layman to understand the mathematical aspect of quantum mechanics but he can acquire a general idea of some of its implications which will enable him to appreciate the significance of these researches for the student of Occultism. He will be able to see at least how all these developments which have bewildered scientists are bringing them nearer to the realization that the purely materialistic philosophy of Science is not adequate and it is worth while examining the Occult doctrine to see whether it can throw some light on their problems. And he will also see the wisdom of holding his judgment in suspense if some of the conclusions at which Science has arrived do not agree with or are even in direct contradiction with the doctrines of Occultism.

The cause of all bewilderment in the scientific world caused by the recent researches in quantum mechanics is the puzzling nature of the quantum, the mysterious " atom of action " which eludes the comprehension of even those who have discovered it. It is well known that in all radiation phenomena, like light, X-rays, gamma-rays etc., energy is emitted or absorbed not continuously but discontinuously, and whatever the nature of the radiation the amount of energy emitted in one of these periods of emission multiplied by the period of emission is always the same and is equal to 6.55×10^{-27} ergs—seconds. This invariable unit of action is called a quantum or " atom of action ". Evidently, all kinds of radiation, as in the case of matter seem to be divided into particles. As we cannot have less amount of an element than an atom so we cannot have less light than one quantum, but whereas there are 92 different kinds of material elements there is only one kind of quantum or " atom of action."

Leaving aside the other interesting facts discovered concerning the nature of the quantum let us refer here briefly only to one point of great significance to the student of Occultism. This point concerns the amount of energy packed in a single quantum. It follows from the nature of the quantum that though it is the same for all kinds of radiations the amount of energy packed in a single quantum will differ according to the wavelength of the radiation. Each quantum carries an amount of energy given by the famous equation of Planck

$$E = nv$$

where v is the frequency of radiation and n is Planck's constant, a very small number—6.6×10^{-27}

It follows from this equation that if the frequency could be large enough or, what is the same thing, if the wavelength could be short enough one single quantum could have all the energy of the universe. To the physicist this is merely an interesting though puzzling conclusion arrived at with regard to the nature of the quantum on the basis of mathematical considerations. But to the student of Occultism this is a fact of great significance and shows how the doctrines of Occultism are being gradually corroborated by the researches of modern Science.

In order to appreciate the significance of the scientific conclusion regarding the relation of the wavelength of the radiation and the amount of energy associated with the radiation we have to remember that when a line is progressively reduced in length it is ultimately reduced to a point and its length under these conditions becomes zero. It follows therefore from the above equation that when the wavelength of a radiation is zero and the radiation takes place through a point an infinite amount of energy can be packed in the radiation. In other words, in the peculiar conditions when there is influx of energy through a point from a higher dimension, an infinite or unlimited amount of energy can be released in the world of lower dimension.

This is exactly the Occult doctrine regarding the source of the tremendous amount of energy which we find in the physical universe and which modern Science cannot account for satisfactorily. The Reality which is present at the heart of a manifested universe has inherent in it an infinite amount of potential power and when this Reality manifests through a point it can release *any* amount of energy needed in the manifested system. It is this energy which in its infinite permutations and combinations forms the objective basis of the universe, and through the agency of the three *Guṇas* produces the impression of an apparently real world in the minds of the Monads who are involved in the manifested system. The law of conservation of energy applies only to the energy which has been released in manifestation through a point, not to its source.

CHAPTER X

FORCE, MIND AND MOTION

WHETHER " Force " as it was conceived by Science in the early stage of its development is a reality or a mere hypothetical assumption to account for the motion of bodies has been a controversial matter from the very beginning in scientific circles. Many scientists regarded the concept of " action-at-a-distance " with scepticism. How could bodies made of matter exert a physical force called gravitation across millions of miles of empty space which is like a vacuum? How could a magnet attract a piece of iron by some kind of mysterious force called magnetism? they thought. But though such doubts continued to trouble scientists for three centuries after Newton, the large majority of them accepted the idea of " action-at-a-distance " by forces because the laws of motion formulated by Newton on the basis of the existence of the force of gravitation accounted exactly for the motion of heavenly bodies.

Einstein was very sceptical about Newton's assumptions and challenged the concept of gravitation in a very definite manner. He rejected the idea of gravitation being a force which can be exerted instantaneously over distances of millions of miles. He evolved a new theory of gravitation in his General Theory of Relativity which is considered by scientists to give a more correct picture of the phenomena in Nature than Newton's concept.

In formulating this theory, Einstein started with the idea which had been developed previously by James Maxwell in connection with the behaviour of magnets. According to Maxwell, a magnet creates a " magnetic field " round it whose structure can be made visible by shaking some iron filings on a piece of paper held above the magnet. The iron filings move and become arranged in accordance with the nature of this field. Maxwell described exactly the structure of such a magnetic field in his famous " field equations " which have formed the basis of all calculations and developments in the field of electro-magnetic engineering during the last century.

Einstein took this idea of the creation of a magnetic field and applied it to gravitational phenomena. According to him, all physical objects produce round them a "gravitational field ", which has a definite structure, and just as the movement of a piece of iron is guided by the structure of the magnetic field so the movement of any body in a gravitational field is determined by the " geometry of that field ". Like Maxwell, Einstein formulated equations for defining the structure of such a gravitational field or the " field properties " of the " space-time continuum ".

It will be seen, therefore, that the concept of gravitation in the Theory of Relativity is quite different from that of Newton. According to this theory " gravitation " is not a " force " at all and Einstein's Law of Gravitation describes the behaviour and movement of objects like planets, etc., not in terms of attractions exerted by heavenly bodies but in terms of the structures of the gravitational fields created in their neighbourhood.

It is not necessary to go further into this aspect of the Theory of Relativity. The object of the above paragraphs was not to discuss this theory. It was to show how the concept of physical " forces " producing various phenomena of nature which held the field for about three centuries has been replaced by another concept which contains nothing about force and is far more complicated. Newton's concept was at least simple and easily comprehensible. Einstein's concept though based upon very complex mathematical equations is utterly incomprehenible to the human mind. But it is the latest and holds the field until it is replaced by another concept probably still more complex and incomprehensible.

One interesting fact which will strike the student of Occultism is how scientists change their principles concerning the ideals and methods of Science as it suits their convenience or immediate purpose. Einstein disbelieved Newton's laws because they appeared improbable and against common sense and formulated his own General Theory of Relativity to account for gravitation. The contempt of Einstein for common sense is graphically expressed in the following quotation from Barnett's book. "But as Einstein has pointed out, common sense is actually nothing more than a deposit of prejudices laid down in the mind prior to the age of eighteen. Every new idea one encounters in later years must combat this accretion of 'self-evident' concepts. And it is because of Einstein's unwillingness ever to accept any unproven principle as self-evident that he was able to penetrate closer to the underlying realities of nature than any scientist before him."

And yet Einstein's Theory bristles with unproven assumptions. If there are no unproven assumptions in his theory and the conclusions are derived mathematically from *proven* facts, why, it may be asked, is there this uncertainty and lack of confidence as reflected in his statement: "No amount of experimentation can ever prove me right, a single experiment may at any time prove me wrong." (Albert Einstein) Why, it may also be asked, use the argument that the superphysical worlds and other realities of the inner worlds cannot exist because the idea is against common sense. They are facts perceived by direct perception by those who, like the Buddha and the Christ, tower even above the greatest scientists the world has ever produced and no one in his senses will ever dream of suspecting that the truths which they have proclaimed unequivocally were the result of either deception or hallucination. If direct perception of truth is to be ruled out as a means of arriving at truth, how can abstract mathematical truths arrived at by scientists be taken as valid, because they are also based ultimately upon mental perception. Even mathematical abstractions cannot stand apart from mental perception though they may be considered independent of sensuous cognition. That is why Indian philosophers have laid down three definite and positive means of arriving at right knowledge as pointed out in aphorism I-7 of the *Yoga-Sūtras*. These are direct cognition, inference and testimony.

The repudiation of the concept of physical force as the cause of motion—uniform or accelerated—in the Theory of Relativity does not mean that occultists accept the concept of Einstein. What has been said above is meant only to show that even well-established scientific theories, if they are opposed to Occult doctrines, do not necessarily mean that the Occult doctrines are incorrect or invalid. As Newton's concept of force of gravitation has been replaced by Einstein's concept of accelerated motion so Einstein's concept can also be replaced by another more ingenious concept and this process of substitution of one theory by another must continue until Science renounces its present methods and the real truth is arrived at by the only method by which it can be arrived at with certainty—direct perception.

The conception embodied in the General Theory is of special interest to the student of the Occult doctrine not because this doctrine considers accelerated motion as the cause of so-called forces but because it does not subscribe to the existence of hidden mysterious " physical " forces which produce all kinds of motions by acting upon physical bodies. By accounting for such forces on the basis of motion Einstein not only eliminated scientific theories based upon such forces but also integrated different kinds of forces like gravity, inertia, cohesion, light, heat, electricity, magnetism, chemical affinity into one single phenomenon of motion.

Although this particular conception does not represent the truth of manifested existence as embodied in the Occult doctrine, still, it takes Science one step nearer to that truth. For, according to the Occult doctrine there is only one Power, Energy or *Shakti* at the basis of the objective universe and all " forces ", motions, etc. are only its differentiated forms. These so-called forces manifest through *Ākāsha* and its derivatives, being initiated and guided by *Chitta* or Mind in its different aspects. It is because *Chitta* and *Ākāsha* or the Divine Mind and Primordial Substance work together from this level in creating the manifested worlds with their subjective and objective phenomena, that this mysterious joint parent of the manifested world is called *Chidākāsha* (*Chitta*+*Ākāsha*) or " Mental Space " in Sanskrit. The relations of all the realities involved in the appearance of the manifest from the unmanifest may be represented as follows:

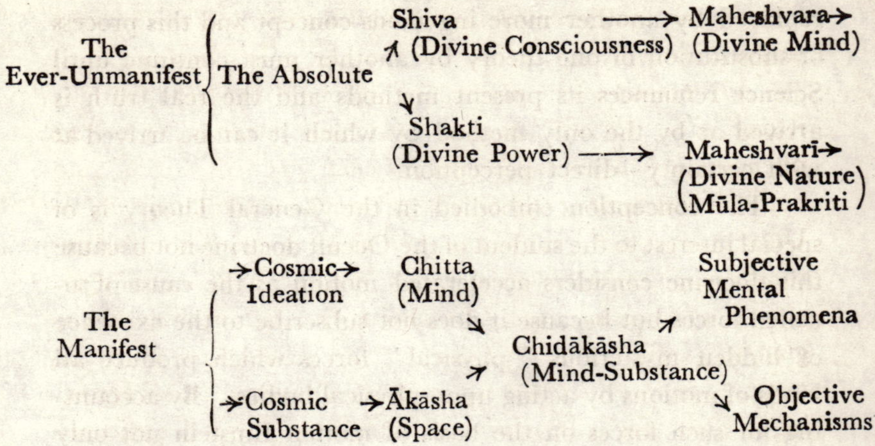

What we should note carefully in the present context is that according to Occultism motion is essentially a function of mind and does not result from unconscious mechanical agencies which work automatically and which we generally indicate by the term "forces". It is true that mind acts through apparently objective "material" agencies in the manifested world but these agencies are themselves derivatives of Mind or rather Mind-Substance called *Chidākāsha* as shown above. All objects will thus be seen to be essentially and completely mental in nature—products of mind in its subjective and objective aspects. That is why movements within them or associated with them as a whole (planetary, electronic etc) are based upon and governed by mind. No independent and mechanical forces, as imagined by us, are involved in the phenomena.

This mental nature of the phenomenal world is implicit in the Occult doctrine according to which all natural phenomena are the result of and guided by "conscious" forces or divine agencies. A purely material universe or partly mental universe motivated by independent insentient mechanical forces could not possibly be amenable to mental guidance of this nature.

If the basis of the manifested worlds is mental, essentially, it can be seen easily how all movements involved in natural phenomena will be produced in such ideated worlds automatically according to the laws which are inherent in the Universal Mind. The suns will shine and the planets rotate and evolve not as a result of blind mechanical forces acting upon them from without, or due to constant Divine guidance from within, but simply because they are essentially mental phenomena taking place in the Universal Mind according to Universal Laws inherent in this Mind. It is the existence of these laws underlying manifestation which accounts for the compulsive force behind all natural phenomena.

The scientific conception of a magnetic or gravitational field being produced and determining the motion of objects will perhaps help us to understand how the presence of the Divine Mind within the manifested universe ensures that all natural phenomena will take place automatically in the universe according to the Universal Laws present in the Universal Mind. The Divine Mind which has created the manifested world and pervades that world can also be considered to create a definite "field" which determines not only the nature and manner of all kinds of movements taking place in the universe but also the nature and manner of all kinds of phenomena in different spheres and at different levels. This unique field created and maintained by the Universal Mind must by its very nature be so comprehensive and all-inclusive that it can provide for the occurrence of all kinds of natural phenomena which are possible in the universe. The "field properties" of this "Divine field" if we may use such a phrase, may be considered to be defined by the totality of Universal Laws which are inherent in the Universal Mind and derived from Divine Consciousness by differentiation.

No one can, of course, define the structure of this field by mathematical equations, as is possible in the case of specific kinds of phenomena like magnetism or gravitation, because the Divine Consciousness is an integrated state and the field which it creates and maintains within the universe must also be an integrated state. For, only in such an "integrated field" can all kinds of natural phenomena take place according to the conditions which are present at any moment. If the specific field created in a particular kind of phenomenon may be called a *Kshetra*—the Sanskrit equivalent of "field"—then this integrated, all-inclusive field may be called *Mahā-kshetra*.

It is this all-inclusive field pervading every manifested system which is the real nature of *Prakriti* the correlate of *Purusha*. The Sanskrit word *Prakriti* means "nature". In this particular context, it obviously means Divine Nature which is co-extensive with Divine Consciousness and pervades the manifested universe like the Divine Consciousness. The phenomena in the manifested universe take place as they do because *Prakriti* is always present everywhere in the background and provides the necessary field or conditions for their occurrence. These conditions are not provided in a haphazard manner but according to natural laws which are inherent in the Universal Mind. These laws are as definite and work with as much exactitude as the laws of gravitation or magnetism, but being far more complex and to a great extent unknown, we cannot calculate or predict their resultant effects in a particular set of conditions. Science has discovered only a few of these laws acting on the surface of the manifested universe while the knowledge of the totality of laws inherent in Divine Nature lies still buried in the vast realm of the unknown.

It will be seen therefore that the Occult conception of natural phenomena though apparently similar to the scientific conception, inasmuch as both are considered due to motion, is fundamentally different. The difference between the two is basically due to the difference in the nature of motion as conceived by Occultism on the one hand and by Science on the other. According to the Theory of Relativity the cause is motion—uniform or accelerated—but motion of a purely physical mechanical nature. According to Occultism also, the cause is motion but this motion is in mental stuff and guided by the Divine Mind, the "field" is mental and is governed by laws present in the Divine Mind and therefore ultimately under the control of the Divine Will. According to the Occult doctrine also there are mechanisms and laws governing these mechanisms but these are again essentially and basically mental in nature.

What is true of gravitation and magnetism is also true of other so-called "forces" working within the atom. These are produced by the mind working on a microcosmic instead of the macrocosmic level. Occultism does not regard the infinitely vast and the infinitesimally small as fundamentally different in nature. Both are relative and merely occupy different places on an infinite scale of magnitudes as has been shown elsewhere.

The Occult doctrine not only presents a far more reasonable, comprehensible, harmonized and beautiful conception of the universe, it is also free from any conclusions which do violence to our common sense and appear utterly unreasonable. Although most scientists have accepted willy-nilly these conclusions to which they have been driven by the inexorable logic of mathematical deduction I am sure at least some of them, in their heart of hearts, have a feeling of revulsion in being obliged to accept and propagate these ideas which appear incomprehensible if not absurd to the common man who has not renounced his common sense. Many of them might be feeling that there is something wrong somewhere in this frightful and meaningless conception of the universe which Science has so laboriously and carefully built up. No doubt, there is something wrong. Where? In the basic assumptions of scientific thought from which they have started in drawing their conclusions. For, facts never lie and mathematics cannot lead to incorrect conclusions if the premises are correct and in accordance with facts.

We need not go further into this question here but it is worth while considering a few facts which show how essentially reasonable and probable the Occult doctrine is in its conception of a mental basis of the manifested universe.

Does a *physical* force always come into play when motions of different kinds are initiated in the phenomenal world? By no means. Let us take the working of a governmental machinery. It works in a certain way because there is will-power behind it. Whose will? That of a king in a kingdom or that of a dictator in a totalitarian regime. In a democracy the will of a single individual is replaced by the collective will of the executive authority which controls the governmental machinery. In the last world war a whole army of millions of soldiers and military officers moved incessantly to carry out the will of Hitler. His will was the direct motivating cause of not one movement but an infinite number of movements of innumerable variety which would not have taken place but for that driving will behind the army. These movements were not only mental but also physical. Did all these movements take place as a result of the action of physical forces? The very idea is absurd. They were all caused by Hitler's will acting through his mind as pointed out above. His will acted upon the minds of people and these minds produced movements in their bodies such as walking, writing, speaking, in carrying out all kinds of tasks.

If motions of *physical* bodies of vast number of individuals can be produced in this visible and obvious manner by the mind and will of a single human individual, why cannot motions of the heavenly bodies be produced by the Divine Will acting through the Divine Mind? According to the Occult doctrine there is mind behind everything from a sun to an atom, and the Divine Will and the Divine Mind acts upon these minds and produces also the physical movements along with the inner mental movements which take place in the invisible worlds. It is true that many kinds of intermediate agencies and mechanisms are involved in this fulfilment of the Divine Will but the fact remains that the ultimate driving power behind them is the Divine Will. This Will does not work in a haphazard or lawless manner. It works generally according to universal laws and plans which are present in the Universal Mind and which are reflected partly in the physical laws which Science has discovered. So, the doctrine of the mental basis of the universe does not mean, as imagined by some people, the repudiation of the scientific knowledge and scientific laws. It merely means looking deeper into the realities of the universe and not merely at the outer shell which the physical universe is really. And this more profound and deeper conception has the tremendous advantage that it is based upon the direct perception of these inner realities in which there is no possibility of error and not mere theories which may or may not be true or valid.

Let us look at the same question from another angle. When the body of a chick is fashioned within an egg or an oak grows out of an acorn no doubt physical and chemical forces are involved but where does the prototype of these organisms come from? Can insentient atoms and physical forces produce such perfect and predetermined forms wherever and whenever a particular kind of germ happens to be present in a particular kind of conditions? Can these very atoms and physical forces produce by themselves a Taj Mahal or an aeroplane without the help of mind? If not, why not?

Another obvious fact which we have to take note of in considering the mental basis of the universe is that mind and consciousness cannot be eliminated in any act of observation, reasoning or drawing of conclusions based upon them. It is really remarkable how easily we can ignore facts which are inconvenient and go against our prejudices and how we can devise reasons and means for ignoring those facts. The whole Theory of Relativity, from beginning to end, is based upon observations, arguments and reasoning involving the human mind and it is this human mind which is sought to be completely eliminated and replaced by a reality of mathematical abstractions existing independently of this mind!

Let the student be careful not to be deceived by the specious argument that orthodox Science does not ignore the mind in formulating its theories, it merely considers the mental processes which are involved as the product of physical forces and reactions between atoms and molecules. The important point at issue is not whether there is a mind or not—only a mad man could deny the existence of mind—but whether this mind is free and independent of the physical organism of the brain through which it normally functions. If it is merely and solely the product of the physical and chemical reactions between the atoms, molecules of the brain, it has no independent existence and reality of its own, and the conclusions arrived at through its agency have no validity and no value because they are the products of interaction of insentient entities without any intelligence.

Do those who are claiming to solve the ultimate riddle of the universe mean to suggest seriously that the processes of reasoning and drawing conclusions which they prize so highly are merely the by-products of the interaction of physical forces acting between atoms and molecules in their brain? If they are, then what is their value? Why should we take conclusions arrived at by the haphazard interactions of atoms and molecules and blind forces of nature seriously and act upon them blindly? The fact really is that there is an amazing lack of clear thinking when scientists set out to solve the riddle of the universe and attempt to foist on a gullible public their grandiose theories about the ultimate problems of existence and human life. If they were a little more open-minded and not obsessed with their prejudices and favourite pursuits, they could themselves see all these inconsistencies and contradictions in their philosophy of scientific materialism.

Unconscious though he is of this fact, the orthodox scientist is poised on the horns of a dilemma. If all mental processes are the result of reactions between atoms, molecules and blind forces of nature, and mind has no independent existence and capacity to act freely, then the scientific conclusions arrived at through such mechanical processes can have no validity or value. Whether the principle of determinism or the principle of uncertainty is true does not make any difference to this argument for both principles are based upon the mechanical character of the underlying processes.

If, on the other hand, the scientist recognizes that human mind has an independent existence and is thus free and able to reason and arrive at intelligent conclusions, based on its own working, then the whole philosophy of scientific materialism falls to the ground. For, if mind and consciousness have an independent existence and reality of their own, an entirely new unknown and powerful factor is introduced in their thinking and reasoning, because, they do not know what realities exist within the subtle and hidden realms of these two. They also do not know what kind of relation exists between mind and matter. In ignoring the human mind so far they have remained completely ignorant of its real nature, possibilities and powers. All their knowledge with regard to it is confined to physical data obtained by experimenting with the physical brain. The whole basis of their philosophy having been rendered invalid by recognizing the independent existence of the mind they will have to start thinking *ab initio*.

It will be seen from the above what a complex problem orthodox scientists have created for themselves by ignoring the basic realities of human life—mind and consciousness—in their search for truth. All the inconsistencies and incomprehensibilities of modern scientific conclusions regarding the larger problems of human life stem from this basic assumption which has been made unnecessarily and arbitrarily in the very beginning, and which has been stuck to doggedly in spite of all the indications to the contrary.

CHAPTER XI

MOTION AND MANIFESTATION

Occultism had long known and declared unequivocally that Energy in the form of motion in mind stuff was the ultimate basis of the whole manifested universe existing at different levels and in different degrees of subtlety. The fundamental nature of Motion and the fact that it can account for many important physical phenomena such as gravitation, electro-magnetism, etc. was realized when Einstein put forward his Theory of Relativity. He doubted that the balance of gravitation and inertia, which according to Newton accounted for the fact that all objects fall at the same rate regardless of their inertial mass, is accidental. He also rejected the idea that gravitation being a force can be exerted instantaneously over great distances. His General Theory is based on the fact that accelerated motion produces the same phenomena as gravitation and it is unnecessary to assume the presence of a force like gravitation to account for the movement of bodies in space.

We need not enter here into the question as to how relative motion can account for universal and fundamental phenomena of Nature. The important point to note is that mere motion can, even according to Science, produce phenomena which are not apparently related to it. The Occult conception of motion and the manner in which it serves as the basis of *all* phenomena of Nature is quite different from the scientific conception as formulated in the General Theory of Relativity. But the scientific theory regarding the role of motion in manifestation can be considered as a step in the direction and partial corroboration of the Occult doctrine. For, if relative motion of material objects can account for a host of physical phenomena, why cannot relative motion of mental images corresponding to those physical objects account for the physical phenomena? For, even according to materialistic philosophy what we perceive is only our own mental image, and Science does not know what is really the nature of the " matter " which constitutes the physical objects and produces these sensuous mental images in our mind.

One of the most striking aspects of the Theory of Relativity which the student of Occultism will notice is how the scientist tries to evade the question of the existence of mind and consciousness as independent principles of existence in manifestation. Even in the cognition and formulation of purely physical phenomena mind comes in everywhere and forms the basis of the whole fabric of scientific knowledge and without the mind working in different ways, in observation, conception and reasoning the whole field of scientific thought will be without any foundation. And yet the orthodox scientist persists in ignoring it and not even acknowledging its role in his work. Why? Because he knows intuitively that once he opens the door for the entry of mind, consciousness and Spirit will follow and make short work of his purely materialistic philosophy and the ideals and values based on this philosophy.

The basic medium in which this motion takes place is *Ākāsha* or rather *Chidākāsha* the " mental space " which is the undifferentiated Substance of the Divine Mind. This *Ākāsha* is one of the most mysterious and significant realities about which Occultism has given out just a hint to the outer world. Being in its highest aspect the abode of the Ultimate Reality and in its lowest aspect the foundation of the objective universe it has naturally many aspects each of which can provide a fascinating and fruitful field for philosophical thought and occult research. But little is known about it and a nebulous idea as the parent of the *Pancha-Bhūtas* is all that is generally present in the mind of the ordinary student concerning *Ākāsha*.

In spite of the doctrine of the *Pancha-Bhūtas* being in the field of Indian Philosophy for thousands of years, the nature of *Ākāsha* has not been fully understood by Indian philosophers. The tremendous advances made in many fields of scientific research have left the thought in this field unaffected. The tendency to take things for granted instead of going deeper into problems is the bane of Indian philosophical thought. Once a thing is stated by any authority all enquiry into it stops and it becomes a sort of counter of thought which is exchanged in discussions without bothering what the counter represents. There are discussions and arguments galore but these are mostly for maintaining certain accepted points of view than for finding the real truth. Inevitably, under these conditions the significance of the thing remains as vague as ever and the doctrine is not enriched by further investigation and enquiry and the addition of real knowledge.

The amount of factual and precise knowledge which has been given out by the Adepts of Occultism about *Ākāsha* is very fragmentary and deliberately veiled. The reason for this lies in the fact that some of the most important secrets of Nature involving the manipulation of tremendously powerful forces are hidden in *Ākāsha*. The possession of this knowledge, which does not require the paraphernalia of any physical instruments in the manipulation of these natural forces, will be positively dangerous for humanity at its present stage of moral and spiritual development. If we want to understand the nature of *Ākāsha* to some extent we shall have to make the attempt under these limitations. There is, however, nothing wrong in doing so provided our motive is to understand the problem from the philosophical and scientific point of view and not to find a short-cut to personal power for selfish purposes.

The Occult knowledge concerning *Ākāsha* may be generally summarized as follows:

Ākāsha, according to the Occult doctrine is the matrix and substratum and thus the origin of all kinds of energies and motions from which all kinds of vibrations and elementary particles are derived. These energies and modes of motion find expression primarily in different kinds of vibrations like sound, light, heat etc., and from these vibrations are then elaborated the different kinds of elementary particles by an infinite variety of permutations and combinations.

This doctrine is in perfect accord with the latest scientific ideas concerning the nature of matter and energy and their mutual relationship. According to the latest discoveries made in the field of physics " matter is nothing but bottled-up radiation ". This agrees with the Occult doctrine that free radiation or vibration is a product of *Ākāsha* and matter is produced when this radiation is bound and confined in the form of elements. A thing has necessarily to be produced in the free state before it can be bottled up.

Let us first dwell for a while on the nature of the vibrations which emanate from *Ākāsha*. *Ākāsha* being an integrated state it can be considered as being capable of giving rise to an infinite number and variety of vibrations, but let us try to understand the nature and role of two of these with which we are concerned in the present context. These two kinds of primary vibrations are of the nature of sound and light, but it is necessary to note that sound and light with which we are familiar and which Science has investigated so thoroughly are merely the lowest and very limited expressions of the sound and light as they emerge from *Ākāsha*.

These two kinds of primary vibrations differ from each other fundamentally in their nature and functions as do their lowest expressions physical sound and physical light. This difference in their essential nature and roles can be seen to some extent in the sphere of sensuous cognition, sound being an instrument of hearing and light that of seeing. But this is not the only or the most important sphere in which their essential difference is demonstrated. It is in the realm of atomic structure that we see this difference in their role brought out in the most remarkable manner. We shall deal with these subjects in detail in other chapters, but let us first take a bird's-eye view of the whole field.

As is well known an atom consists of two parts, the outer sphere consisting of shells formed by the whirling electrons and the nucleus at the centre consisting of protons and neutrons. Science has made some of its most brilliant discoveries in unravelling the structure of the atom. But though it has been able to obtain a fairly clear picture of the outer sphere of the atom consisting of electronic shells, its knowledge concerning the nucleus is very hazy and uncertain. It knows that the nucleus consists of positively charged protons and electrically neutral neutrons, the total positive charge on the protons being equal to the total negative charge on the electrons, and the atom as a whole being electrically neutral. But there are many facts about the nucleus which are inexplicable and a complete mystery to the atomic scientist. For example, how are the large number of positively charged protons which are present in the nucleus of heavy elements and which must exert a tremendous repelling force towards one another, held together? What is it that binds together these mutually repelling particles?

It is true that they have made tentative assumptions with regard to the nature of these forces and even measured them in physical terms. They have also given them names. But these are mere words which do not give any indication of the nature of the forces which are acting within the nucleus of the atom. So the statement that matter is bottled up radiation enunciates merely a general principle and does not give any real indication as to how radiation has been bottled up and how it exists within the primary units of matter, namely atoms.

Occultism does not give us a detailed picture of the structure of the atom in these respects but it gives us an overall picture which is comprehensible and in harmony with the Occult doctrine as a whole. It does not believe in entering into and getting so much involved in details that people " cannot see the wood for the trees ". Besides, Occultism regards all knowledge obtained through the intellect incomplete, uncertain, defective and without any relation to the whole Truth of which it is an insignificant part. It, therefore, places before man the ideal of finding that Truth " knowing which everything is known truly, completely and in the correct perspective and man is able to rise above the illusions and limitations of the lower life ". Ordinary knowledge is not despised or ignored but it is considered merely as a stepping stone to the acquisition of the knowledge of Reality and therefore to be pursued and acquired with discrimination. This will show the wide gap which separates the attitude of the occultist and the scientist with regard to the pursuit of knowledge concerning phenomenal things and the wisdom of concentrating only on the essential and general aspects of these things when we are engaged in the pursuit of Real knowledge.

In order to understand the Occult conception concerning the general structure of the atom, it is necessary to be quite clear in one's mind that Occultism does not regard anything in the universe, from an atom to a solar system, as " dead " and insentient. It regards the whole universe as an expression of a Reality which not only pervades and expresses Itself through everything but is the inner and hidden directing and controlling force of every organized unit in manifestation expressed in the well known Occult maxim:

"*Shiva* and *Shakti* are the innermost soul of the Universe ".

Shiva and *Shakti* denote here the two Principles of Divine Consciousness and Divine Power which are polar and inseparable and together constitute the Reality underlying the universe.

The atom, according to the general principle enunciated in the above maxim is the joint expression of the Consciousness of *Shiva* and the Power of *Shakti*. Divine Consciousness is the source of the static centripetal and centrifugal forces within the atom which control and coordinate the different processes and hold it together. Divine Power expresses itself in the motion of the electrons which whirl round the central nucleus and constitutes the kinetic energy content of the atom. Basically, an atom is the result of the interplay of consciousness and power but at this level consciousness expresses itself and appears as a central static force and power expresses itself and appears as the dynamic energy of the revolving electrons. How these forces and energies are coordinated and work together is for Science to work out in detail, but it will never be able to do that satisfactorily and gain a correct picture of the inner working of the atom unless it takes the Occult conception of the atom into account.

The above idea may appear fantastic to scientists and it is not likely to be considered by them seriously at present in view of their antipathy to anything which suggests in any way the existence of a Divine Reality hidden within the universe. But many scientists have begun to sense the existence of a mysterious over-all Intelligence in many organic processes and vital phenomena though they have not the courage to acknowledge this heterodox feeling publicly. At least these people should be able to see that if there can be directive intelligence (which means mind and consciousness) behind biological processes there is no reason why there cannot be some kind of mind and consciousness behind chemical and physical processes also. What justification is there for drawing the line at one particular level of natural phenomena?

And if the existence of mind and consciousness is conceded in the case of atoms and molecules also, then the Occult conception is recognized in principle and it does not matter if this over-all Intelligence is not considered as an expression of Consciousness or Divinity. Science has had to recognize in principle many Occult doctrines, which were previously considered fantastic, owing to the logic of its own discoveries. Let us give it time to make further discoveries, to get further involved in contradictions and confusions, so that it has to turn willy-nilly towards the portals of the Divine Wisdom for light. After all, it is also engaged in the pursuit of truth, and if what Occultism asserts about the basis of the physical universe is true, Science must come to that truth sooner or later. We have to wait and let things develop naturally according to the Divine Plan.

According to the Occult doctrine the atom is not the only unit of organized forces which involves the interplay of Consciousness and Power that is perfectly coordinated and balanced. All such units from an atom to a solar system are governed by the same principles working at different levels and adapted to the nature of the particular manifestation. For, the whole universe at every level and in every sphere is an expression of Divine Consciousness and this is at the basis of every organism which requires intelligent coordination of forces and regulation of natural processes. The coordination and direction of cohesive forces is a function of Consciousness which is carried out through the agency of " Sound ". Providing the structural material is a function of Power which is carried out through the agency of " Light " or radiation as it is called in scientific terminology.

Take the human body. The body is an organism in which innumerable natural processes are being carried on with such perfect coordination and exactitude that it should make any intelligent person not only wonder but ask how this coordination is brought about. I am sure many scientists wonder, but do they ask the question how the activities of the insentient atoms, molecules and cells working in different organs of the human body are coordinated with such perfection without there being an over-all intelligence of a high order behind these natural processes? And if there is such an intelligence, what is its source and where does it reside?

According to the Occult doctrine there is an all-pervading Divine mind and consciousness behind all such organisms which controls, coordinates and regulates all these processes and this control, coordination and regulation is effected through the agency of " Sound " working through the *Prānamaya Kosha* or the " etheric double ". This inner " sound ", which is normally inaudible, proceeds from the "heart" which is the seat of the "*Ātmā*" and it is in the field created by the "Sound" that the body functions.

This sound is called *Anāhata Shabda* "the eternal sound produced without vibration of a material instrument". A discipline involving the hearing of this sound is the basis of some systems of *Laya Yoga* in which the mind is merged in consciousness through the agency of this sound. According to the Upanishads this "sound" in the heart ceases to be heard some time before the death of the physical body and such cessation is taken as a premonition of impending death. This fact is significant because it shows that the activities of the body are controlled and regulated by "sound" proceeding from the "centre" of the body.

The highest and most spectacular physical expression of the Divine Consciousness and Power is, of course, a solar system. According to Science the sun is merely an incandescent material body burning its own matter to supply light, heat and other forms of energy to the solar system and doomed to "death" when the energy stored within it is spent. But according to Occultism it is the physical abode and instrument of a Divine Being—called a Solar Logos or *Īshvara*—whose consciousness pervades the solar system, controls and regulates all its activities in every sphere at all levels and provides the field for the evolution of innumerable Monads involved in the particular manifestation. Here again "sound" plays a fundamental role in the coordination and regulation of all activities and natural processes and "Light" in the structural basis of the manifested solar system, "sound" being the instrument of the static positive principle of Consciousness and light that of the dynamic, negative principle of Power.

Another aspect of motion as a basis of manifestation is shown in the doctrine of the three *Guṇas*. This Occult doctrine has been discussed in some detail in *Man, God and the Universe* but a brief review of the doctrine may be given here to provide an all-round general idea of the subject of this article.

As pointed out above, according to the Occult doctrine manifestation is essentially an interplay of Consciousness and Power, the two primary and polar principles arising out of the non-differentiated One Reality. In the realm of manifestation with which alone Sānkhya deals these two Principles are called *Purusha* and *Prakriti*, the *Purusha* being considered the subjective element in manifestation and *Prakriti* not only the objective mechanism, but also the controller and regulator of the objective mechanism. Dealing as it did only with the state of manifestation and considering *Purusha* merely as a witness Sānkhya had to attribute not only the role of providing the objective mechanism but also that of controlling and running it to *Prakriti* although this has made this sytem of philosophy appear self-contradictory and incomplete. The role of controlling and running the objective mechanism from within belongs really to *Shiva* although these functions being in the realm of the Unmanifest are incomprehensible to the intellect. In the same way the role of providing the mechanism belongs really to *Shakti* which is also beyond manifestation and therefore outside the purview of Sānkhya philosophy.

As we are discussing motion in relation to manifestation we are concerned here only with the objective mechanism in the sense in which the scientist uses the word Nature. It is this objective mechanism which is based on motion and motion only.

This motion according to the doctrine of the three *Guṇas* is of three kinds (1) Rhythmic motion (2) Non-Rhythmic motion and (3) No-Motion. As has been shown in another context, these are the only three kinds of motion which are possible, if we consider No-motion or stability also as an aspect of motion.

The word *Guṇa* in Sanskrit means "property" in its essential nature which is motion. According to the doctrine of the three *Guṇas* there are only three kinds of fundamental motions from which all kinds of properties are derived just as all possible colours are derived from the three primary colours, red, yellow and blue. The most important point to note in considering the doctrine of the three *Guṇas* is that all properties depend upon the constitution of the constituent particles and this constitution depends upon the particular combination of the three kinds of motion mentioned above. So, the infinite variety of properties are derived from and related to the infinite variety of permutations and combinations of the three kinds of motion, each to each.

That the properties of an object or substance cognized by the sense-organs depend upon the constitution of the particles which compose the object or substance is a scientific fact. That this constitution of the particles depends upon the particular permutation and combination of the three types of motion is also a scientific fact. So the doctrine of the three *Guṇas* has been broadly and amply corroborated by researches in the field of modern Science.

It will be seen from what has been said above that motion, in its three forms, can not only be elaborated into all kinds of combinations which constitute atoms and molecules and form the structural basis of the universe but each combination corresponds to a subjective state of the mind which is the essential nature of sensation. Sensation which is called *Tanmātrā* is thus the subjective or mental counterpart of the objects in the objective environment. It appears in the mind through the instrumentality of the sense-organs or *Indriyas* when the *Bhūtas*, the five cosmic principles which stimulate sensations, act upon the mind through the instrumentality of atoms and molecules and vibrations emanating from them. Each sensation thus corresponds exactly to a particular combination of motion and it is in this manner that the subjective and objective worlds are mathematically related.

The functioning of the One Reality in its subjective and objective aspects as It descends to the lowest planes of manifestation and these two aspects which interact with each other on the physical plane may be represented as follows:

THE ONE REALITY

↙		↘
Brahma Chaitanya		Shabda Brahman
↓		↓
Consciousness		Power
↓		↓
Mind		Nature
↓		↓
Sensation		Five Elemental Principles
↓		↓
Sense-Organs	⟶ ⟵	Objects

(*Both in the realm of matter*)

CHAPTER XII

POINTS IN MOTION AS THE ULTIMATE BASIS OF THE OBJECTIVE UNIVERSE

According to Occultism the whole of the manifested universe is atomic in its nature, whether on the physical or superphysical planes. In the realm of matter, the atoms become smaller and smaller as we penetrate inwards into the superphysical planes of Nature. Where does this infinitesimal nature of the atom end, or in other words, what is the smallest size of the atom beyond which we cannot go further? Obviously, the ideal point which is the limit of atomicity or smallness.

So, the ultimate basis of the material world is not really atomic but "pointed" if we may put it this way. The material world, in all its degrees of subtlety and complexity, is really at its basis a combination of an infinite number of ideal points. It is from these points that the atoms of the different planes are elaborated, step by step, in greater and greater complexity. It is these points which by their different movements and combinations can produce all the properties which we cognize with our sense-organs. Our ordinary experiences as well as the discoveries made in different fields of Science confirm this Occult doctrine at least partly. We have only to examine a photogravure in any newspaper or journal with a magnifying glass to realize how a number of points arranged in a certain manner can produce the visual impression of any photograph.

We have only to watch a display of fire-works at night to see how moving points of light can produce the impressions of all kinds of figures on our eyes.

Where and in what are these ideal points, which form the ultimate basis of the so-called material world, situated or centred? In Consciousness. The objects of our physical plane can be resolved into atoms of matter and these atoms can be resolved, step by step, into atoms of the subtler planes, hidden one within the other, until we reach the bedrock of the manifested universe and find it to be nothing but a vast concourse of ideal points situated or embedded in consciousness. This is inherent in the very nature of the relation of the unmanifest Reality and the manifest universe which is derived from that Reality. Consciousness can function and express itself only through a point as has been explained in other contexts.

According to the Occult doctrine the whole manifested universe in all its grades of subtlety has appeared through the *Mahābindu* or the Great Point and this Point, like an archetype, is capable of multiplying itself in an infinite number of ideal points. As has been explained elsewhere the prefix *Mahā* in Sanskrit generally denotes a principle or *Tattva* which can serve as an inexhaustible source of an infinite number and variety of expressions of that principle. The ideal points which lie at the basis of the manifested universe may thus be considered as infinite expressions of the *Mahābindu* or the One Reality in its objective aspect. These points mirror the One Reality in an increasing measure, as consciousness which finds expression through them unfolds by means of the evolutionary processes. They are thus the primary vehicles of consciousness. The elaboration of a complex vehicle with name and form is only a transitional device adopted in the intermediate stages of the evolutionary process. The two aspects of the Absolute as the Point and the Void will also be seen in a new light from what has been said above. Their relation corresponds to that of Consciousness and its vehicle on the planes of manifestation.

The student will also be able to get a glimpse into the vast and mysterious cycle of evolution from what has been said above. These ideal points which appear from the *Mahābindu* serve as a vehicle of atoms which evolve through the different planes of the universe and different kingdoms of life and emerge ultimately as conscious *Jīvātmās* on the physical plane. At this stage a new impetus is given to the evolutionary process, owing to the direct association of the Divine Principle in human evolution, and man starts on that unlimited expansion and unfoldment of consciousness which first makes him a super-human being and then a Solar Logos. The Logoi also continue to evolve and draw nearer and nearer to the Cosmic Logos, the first and the highest expression of the Ultimate Reality on the planes of manifestation. In this way is accomplished both the involution and evolution of Consciousness and we are able to see the whole cycle of the evolutionary process at a glance, the Monad appearing from the Ever-Unmanifest apparently as a point-atom and again disappearing into the Ever-Unmanifest as a Logos. What happens in that dark realm of the Ever-Unmanifest, who can say?

It is this cyclic and eternal process of involution and evolution of consciousness which is represented in the symbol of the snake swallowing its tail. The fact that this sacred symbol has been included in the seal of the Theosophical Society shows how important this Occult doctrine was considered by the founders of the Theosophical Society. Of course, from the highest point of view the whole process can be considered as an illusion. For, it is an interpretation in terms of Time and Space of some Reality which exists in the Unmanifest. Time and Space themselves being illusory in nature, an interpretation based upon them cannot represent the Real Truth. This limitation applies not only to this particular doctrine but to all knowledge in the realm of the intellect. As long as we are imprisoned within our intellect we have to be satisfied with what the intellect can do to enable us to understand the mysteries of the Real world which exists within us. For knowing the Real Truth we must go beyond the mind.

We have so far considered these points in their static aspect. But manifestation is a dynamic phenomenon and the points which form the basis of manifestation must therefore have a dynamic aspect also. According to the Occult doctrine the whole manifested universe, in its objective aspect, is vibrant, alive and constantly moving. Even those things which appear dead and static apparently are internally in constant motion and possess tremendous amounts of energy locked up within them. This ancient doctrine of Occultism has been corroborated in a spectacular manner by the researches of modern Science. Within the apparently lifeless matter, Science has found molecules in constant motion. Within the molecules we find vibrations of atoms. Within the atoms electrons whirl round the nucleus with inconceivable speeds. Science has not been able to determine the real nature of the nucleus but the principle of equivalence of mass and energy shows conclusively that even the nucleus must be an expression of a tremendous amount of energy locked up in some form. It is this energy which is released when the nucleus is broken up in an uncontrolled manner in an atomic bomb and in a controlled manner in thermonuclear devices.

According to the Occult doctrine the atoms of a lower plane are formed by setting up whirlpools of tremendous energy in the matter of the next higher planes, the atoms thus being essentially an energy phenomenon or phenomenon of motion. It can be seen easily that if the atoms of the physical plane are resolved back into the atoms of the next higher plane, step by step, we shall be left ultimately with nothing but ideal points associated with motion. In other words the manifested universe is, at its ultimate basis, nothing but a phenomenon of moving points or ideal centres in consciousness endowed with energy. It is from these ideal points in motion that the atoms are elaborated, step by step, in their infinite variety by different permutations and combinations. The arrangements and movements of such points can account for all the phenomena of sense-cognition even on the basis of scientific knowledge concerning the structure of the atoms of the physical plane. We cannot go and need not go further into the details of this complicated question considering the tremendous limitations under which consciousness and mind have to function on the lowest plane of the universe—the physical. We have to remain content with a general idea of the underlying principles of manifestation.

Let us now examine in a general way the occult significance of the conception of the manifested universe being nothing but an infinite number of points in motion at its ultimate basis. It has been pointed out in another context that this manifested universe is based upon and is derived from the *Shiva-Shakti Tattva* as hinted at in the well-known maxim:

Shiva-shaktyātmakam Vishvam

" *Shiva* and *Shakti* are the innermost soul of the universe ". If it is derived from these two ultimate Principles—Consciousness and Power—and is the result of their interplay, it follows that the first product or derivation of their interplay must embody within itself the essential nature of both, and this is exactly what a moving ideal point does. As has been shown elsewhere a point can embody within itself unlimited potential power and can also serve as a vehicle of the highest and lowest level of consciousness. As the objective universe is a vast concourse of moving points at its base and all the infinite variety of objects which we perceive on different planes of the universe are composed of different permutations and combinations of these moving points it will be seen how true the Occult maxim referred to above is.

One important conclusion which follows from this fundamental relation of the manifested universe in its objective aspect with the *Shiva-Shakti Tattva* is that it must be completely amenable to the Divine Will embodied in the *Shiva Tattva*. For, if each of the particles which constitute the foundation of the objective universe is an embodiment of the *Shiva-Shakti Tattva* and if the whole of this universe in all its complexity is elaborated by the permutations and combinations of these particles then the Divine Will must be able to control and guide not only these fundamental particles but also all their derivatives on the different planes of the universe. In other words, the Divine Will must be supreme in the realm of the manifested universe in every respect. The capacity in the Divine Consciousness to control and manipulate the objective universe is thus inherent in the very nature of the universe and is not an arbitrary power assumed by the Divine in dealing with the manifested universe.

That matter can be completely controlled and manipulated by consciousness from within, in contradistinction to its partial control and manipulation by scientists from without by external means, is quite clear from the relevant aphorisms on

Bhūta-Jaya (III-45, III-46, III-47) and I-40 in the *Yoga-Sūtras*. But in order to control it in this way it is necessary to be able to penetrate into consciousness to the level of the Divine Mind in which matter is rooted. The five different states of the *Bhūtas* mentioned in aphorism III-45 represent penetration into these different levels of the mind in order to arrive at the bedrock of consciousness from which matter takes its rise.

It will be seen therefore that though the control and manipulation of matter from within is feasible this can be accomplished only by Adepts of Occultism who can control the sources of matter and can utilize the laws of Nature governing the formation and dissolution of matter on the innermost planes of manifestation. No one should imagine that he can accomplish these things by simply concentrating the mind on what has to be done.

But though the manipulation of matter in general is possible only for Adepts of Occultism its partial manipulation in a vital organism like the human body is comparatively easy and many changes can be brought about in the body by simple but effective concentration of mind. The reason for this lies in the fact that the metabolic processes in the body are governed and regulated by *Prāṇa* coursing through the *Prāṇamaya Kosha* (or the Etheric Double) and *Prāṇa* is amenable to the control of the mind. This is the secret of many systems of cure based on the use of thought power.

Another important conclusion which we can draw from the relation of the underlying *Shiva-Shakti Tattva* with the manifested universe is that the Divine Will embodied in the *Shiva Tattva* is not only supreme in the realm of the manifested universe and can control and manipulate all its processes from within, but that the Divine Power embodied in the

Shakti Tattva is infinite and inexhaustible and can supply any amount of energy needed for creating and running the machinery of a manifested universe.

It should be remembered that power is defined in Science as work divided by time and work as force multiplied by distance. A force does not do any work unless it moves. If a grandfather clock is wound up but not started, the weight exerts a force, a pull on the string it hangs on, but while it hangs it does no *work*. Once the clock is set going, the weight descends and turns the machinery, so doing work in overcoming the friction of the moving parts. Similarly, power is exerted only when work is being done and where there is no movement and therefore no work is being done, power is potential. It is considered potential and not non-existent because every system for doing work has a certain limited capacity for doing work and the limit of this capacity is a measure of the potential power which it contains.

Let us now apply these scientific ideas to the relation of the *Shiva-Shakti Tattva* with a manifested system. We see in the first place that the *Shakti Tattva* exists in the Ever-unmanifest and from the very nature of the Unmanifest state the Divine Power in this state must be potential. For, unless there is a manifested system there is no activity, and so no work can be done and where no work is being done, the Power must be necessarily potential.

But, as we have seen in the previous paragraph, though power is potential in any system it has a limiting value for that system. The grandfather clock is capable of doing work but its capacity for doing work in running the machinery of the clock is very small. On the other hand, the booster which raises a spacecraft into orbit has to do a tremendous amount of work and must have correspondingly great power. Now let us try to imagine the amount of work which has to be done in a manifested universe and the amount of energy which is needed for doing this work. Obviously, the Divine Power in the Unmanifest which is the only source of energy must be tremendous.

According to the Occult doctrine this power is not only tremendous but infinite. There is a world of difference between being tremendous and infinite. The former though great has a limit, but the latter has no limit. The former is an attribute of the manifest state and must therefore be in the realm of the manifest state. The latter, on the other hand, is an attribute of the Unmanifest and can exist only in the realm of the Unmanifest. The energy required and the power needed for creating and running the machinery of a solar system is obviously tremendous but the energy and power needed for creating and running the machinery of the universe is infinitely greater. But both, however great, must be limited because they are in the realm of manifestation. The Divine Power in the realm of the Unmanifest, on the other hand, is not only potential but infinite. It is capable of putting forth *any* amount of energy needed for creating and running the machinery of the manifested universe, and not only this present universe but an infinite number of universes which will continue to appear eternally in the alternation of *Srishṭi* and *Pralaya*.

It is this fact which is sought to be conveyed by the Occult maxim:

अनन्तशक्तिखचितं ब्रह्मसर्वेश्वरेश्वरम् ।

Ananta Shakti khachitam Brahma Sarveshvareshvaram

"The Unmanifest Reality or the Supreme Consciousness of *Shiva* who is the hidden Ruler of all manifested

Logoi is endowed with infinite and inexhaustible power or *Shakti* ".

That this statement is based not on any speculative thought in the realm of philosophy or hyperbolic statements frequently found in religious literature but on facts of existence will be clear when we study the recent developments in physics bearing on this question. According to these researches, if the wave-length could be short enough *one single quantum* could have all the energy present in the universe as has been explained in another context.

We have seen that the *Shiva Tattva* is the seat of the infinite Divine Will which can control and manipulate the universe from within. And the *Shakti Tattva* represents the infinite Divine Power which can supply all the energy requirements of any manifested system. Will and Power are correlates and infinite Will must necessarily go with infinite Power. But the existence of infinite Will and infinite Power in the *Shiva-Shakti Tattva* should not lead us to imagine that this state must be characterized by a tremendous amount of tension and potentiality for disturbance. On the contrary, it is a state of utter harmony which cannot be disturbed under any circumstances and of a Peace which is of such a transcendental nature that it is inconceivable in the realm of manifestation. A faint glimmer of its nature can perhaps be obtained only in the " peace which passeth understanding " which we experience on the spiritual planes.

Most people have generally a very wrong notion of will-power. They confuse it with impulsive desire which is always accompanied by tension and instability, apparent or hidden. Spiritual will and in an infinitely greater degree Divine Will is utterly free from tension and instability. The secret of this lies in the fact that tension and instability are always due to lack of confidence and lack of confidence arises out of insufficient power. It is when we do not possess sufficient power or capacity to accomplish what we want, or what we will to do, that we feel lack of self-confidence and this results in tension and disturbance of the mind. The greater our control of the outer situation and the inner state of our mind the more peaceful and calm we can remain under the circumstances. The *Shiva-Shakti Tattva* being the seat of infinite Will and infinite Power must therefore necessarily be a state which is quite free from disturbance or even the possibility of disturbance.

There is another misconception about the Divine Will and Power which we must guard against. The existence of infinite Will and Power at the back of the process of manifestation does not mean that the Logos of a system has to direct and supervise the work of running the gigantic machine in detail all the time. The Divine Plan and the method of implementing that Plan are embodied and inherent in the laws of Nature which underlie all processes of manifestation and therefore the system runs itself and the Plan fulfils itself naturally and automatically without any interference from outside. Even in the case of a microcosm like the human body we can see this automatic performance of all natural processes and achievement of different objectives without any need of conscious action or supervision from the individual who inhabits and uses the body. But when anything goes wrong and needs personal attention the individual can and does intervene to put things right.

Similarly, the machine of a manifested system runs itself and works for the fulfilment of the Divine Plan which is inherent in it, but when anything goes wrong or a new impulse is needed in the evolutionary process Divine intervention does take place. It is true that this intervention comes from within and is carried out through the agents and means which are available, but all the same, it means the introduction of a new and extraordinary force and impulse in the stream of evolution which does not have its origin in the ordinary natural processes. This idea of Divine intervention in such circumstances is embodied in the Occult doctrine of "*Avatāras*" which means "the descent of the Divine from above". It is true that this idea is interpreted in Hinduism as the descent of Divinity in a definite form but this is only to make it easier to understand by the common man. This intervention is not confined to the major crises in the progress of humanity. It can also take place in the life of individuals in response to intense aspiration or a cry for help.

Is this kind of Divine intervention possible according to the scientific concept of the universe? Scientists of the 19th century and early part of this century laughed at this idea because, according to the mechanistic conception of the universe which was prevalent in scientific circles at that time a rigid determinism governed not only the processes of nature but also the life of man. But all this has changed owing to the developments which have taken place in physics in recent years. The discovery of the "principle of uncertainty" has exploded the concept of rigid determinism and shown that both in natural phenomena and the life of man there is room for the exercise of free will.

CHAPTER XIII

THE UNDERLYING MECHANISM OF MANIFESTATION

As has been pointed out elsewhere the knowledge referred to as Occultism is based upon a total view of the universe (or rather the Cosmos) and is not like scientific knowledge which is confined to a very limited field and has been obtained by very limited methods of investigation. Modern Science has assumed arbitrarily and conveniently that there is nothing in existence except the physical world and therefore investigation of physical phenomena by purely physical methods is not only adequate but all that is possible.

The Adepts of Occultism who wanted to investigate the facts of existence assumed nothing and were determined to investigate all that there was to investigate by all possible methods that were available. It will be seen, therefore, that they were the real seekers after Truth and qualified to find that ultimate Truth which lies at the basis of the universe. Scientists, in spite of all their professions, cannot be considered as seekers after Truth by any means. Those who are afraid to face facts of existence, make convenient assumptions without justification, confine themselves deliberately to the attainment of very limited objectives by very limited means, who do not even believe really that there is an ultimate Truth underlying the universe, cannot claim to be seekers after Truth. At best, they can be considered as seekers after the truths underlying the physical world for satisfying their intellectual curiosity and for exploiting the forces of Nature for alleviating suffering and promoting human welfare on the one hand and for inflicting suffering and planning destruction on a mass scale on the other.

As the universe is based upon a Reality which is essentially of the nature of Mind and Consciousness a true and total knowledge of it can be obtained only by going into the realms of mind and consciousness, and by penetrating into their innermost depths for knowing the ultimate Truth of existence. This is not merely a theoretical possibility but a fact of direct experience for the Adepts of Occultism, and anyone who follows in their footsteps can gain the same experience for himself. The doctrines of Occultism are based upon progressive experiences in the realm of mind and consciousness and the final experience of the ultimate Truth in which all truths of existence are contained in their essential, true and harmonized form.

According to the Occult doctrine the underlying Reality from which the universe is derived exists in two forms, the Manifest and the Unmanifest. The Manifest appears from the Unmanifest periodically and after functioning for some time again merges in the Unmanifest, this cycle being repeated eternally in the alternation of *Srishṭi* and *Pralaya.* This process of appearance and disappearance of the Manifest from the Unmanifest takes place in a definite manner and through a definite mechanism. This mechanism will naturally operate at the boundary between the Unmanifest and the Manifest and can therefore be known only by those who are able to transcend the realm of the Manifest and function in the realm of the Unmanifest. This process and the mechanism through which it acts are referred to as *Krama* in the *Yoga-Sūtras* and is referred to in aphorisms III-53, IV-32 and IV-33. From these aphorisms the student will be able to see at least that this *Krama* lies at the very basis of the manifested universe and it is possible to know it in the real sense only by going beyond the realm of manifestation.

On account of the transcendental nature of the knowledge regarding this mechanism it is not possible to comprehend or formulate it intellectually. Still, on the basis of the doctrines of Occultism which have been given out, it is perhaps possible to conceive it at least in shadowy outline. Let us therefore deal with the essential features of this mechanism in the most general manner, as best as we can.

It will perhaps be convenient to start with the Tāntrik doctrine of *Nāda*, *Bindu* and *Kalā* with which most students of Occultism are familiar at least superficially. These three words generally occur together and in every context they are used in connection with transcendence of the world of manifestation and the functioning of consciousness in the world of Reality. This is so because they are related to, and constitute three of the essential aspects of, the mechanism which connects the Unmanifest with the Manifest.

Before we deal with the nature of this mechanism let us first consider briefly the meaning of the three Sanskrit words—*Nāda*, *Bindu*, *Kalā*. *Nāda* ordinarily means sound but in such special contexts in Occult literature the word stands for that mystic " sound " which contains in an integrated form all kinds of " Sounds " which exist and play different roles in the manifested universe. These " sounds " are of many kinds, some of which are quite unknown to us. The sound with which we are familiar on the physical plane and which Science has been able to investigate thoroughly is only one expression of " Sound " as it exists in its totality and integrated form. *Nāda* according to the Occult doctrine is the most fundamental and mysterious Force in manifestation and the source of many Occult powers of the lowest and the highest kinds.

Bindu means a point. In the context of the mechanism of manifestation it means the mystic point from and through which any unmanifest reality is projected into the realm of manifestation. As has been explained in *Man, God and the Universe* the ideal mathematical point has many extraordinary properties which make it an ideal instrument of manifestation and a means of communication between the manifest and the unmanifest. The Sanskrit word *Bindu* also means a " drop ", i.e., any spherical mass of matter held together by central forces. In this sense also it fits into the Occult conception of the mechanism of manifestation, for, every manifested system is such an organism.

Kalā is a Sanskrit word in common use but the special significance of the word when it is used in relation to the mechanism of manifestation is very rarely understood because none of the ordinary connotations of the word fits in properly in this context. We shall deal with this subtle significance later on but it may be pointed out here that *Kalā* stands for the differentiated forms or functions of any principle in the manifest, the principle existing in an integrated form in the unmanifest as *Shabda* or unmanifest " Sound ".

In order to understand the functions and relation of *Nāda, Bindu, Kalā* in the process of manifestation it will be helpful if we recall first the analogous phenomenon of the dispersion of white light into a spectrum of coloured lights by means of a prism. White light is an integrated state containing all lights of the spectrum within it in a potential and mysterious form. When it is passed through a glass prism it is dispersed into coloured lights which form a continuous spectrum on a white surface as has been explained in other contexts. The important point to note in this phenomenon is that it is the peculiar mechanism of the prism which brings about the differentiation of white light and its appearance in the form of coloured lights. The experiment thus provides a very simple but apt illustration of the differentiation of an integrated state into its differentiated states, both of which can be seen present simultaneously.

It has been explained already that a Point or *Bindu* is the instrument of manifestation and wherever an unmanifest state is changed into a manifested state a Point is involved. The function of the Point or *Bindu* in manifestation is thus similar to that of the prism in the dispersion of white light into coloured lights. It differentiates a *Tattva* or a principle in an integrated state into its differentiated forms or functions. This differentiation can be represented diagrammatically as follows:

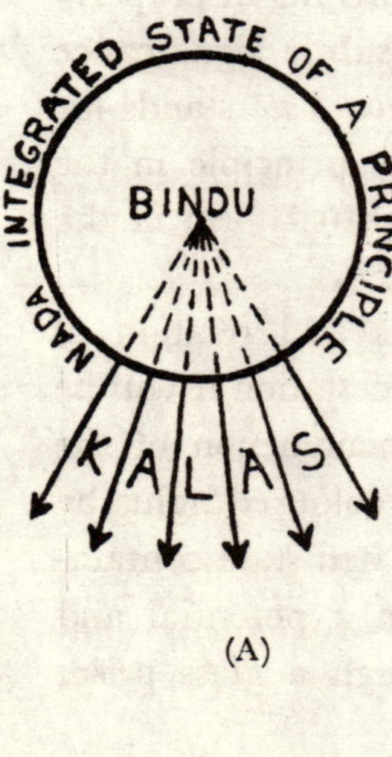

(A)

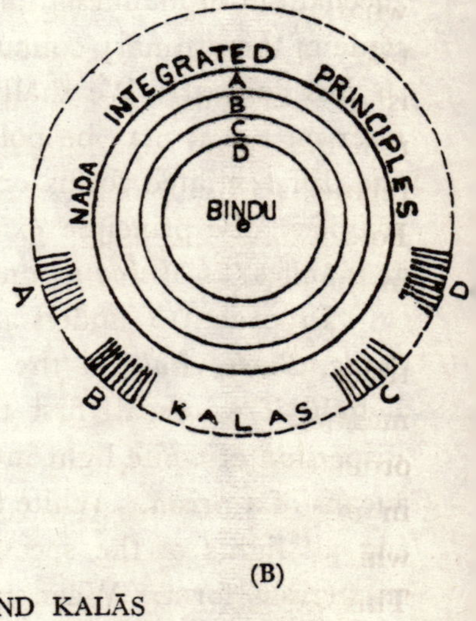

(B)

NĀDA, BINDU AND KALĀS

It will be seen, therefore, that a Point is not only the point of contact or the door between the unmanifest and the manifest but an instrument of dispersing an integrated state of a principle into its corresponding differentiated states in the form of functions. Any principle when it descends from its integrated state in the unmanifest into the differentiated states in the manifest must pass through the Point or the common Centre of Consciousness. All principles in the unmanifest represented by concentric circles may thus be visualized as being projected through the Point into their respective spectrums of differentiated functions or *Kalās* as shown in Figure (B). Each principle is related to its respective spectrum of differentiated *Kalās* and can perform its functions in the realm of manifestation as and when required. It will be seen from what has been said above how closely the phenomenon of light resembles the reality of consciousness.

Let us now consider the other two constituents of the mechanism of manifestation a little more in detail and ask what is the essential nature of *Nāda* and *Kalā*. Let us first take *Nāda*. In order to understand how an integrated principle is differentiated into its respective constituents we have to remember that all principles have two aspects, one static and related to Consciousness, the other dynamic and related to Power. As pure functions they are related to Consciousness, in the expression of those functions they are related to Power.

It is the Supreme Consciousness containing these functions in their totality and highest integrated form in the Unmanifest which is referred to as *Brahma Chaitanya*. But in order that these functions may be capable of being expressed in manifestation they must also be present in another form which can lend itself automatically for their expression. This aspect of Reality in the Unmanifest is referred to as

Shabda Brahma. It is referred to as "*Brahma* as Sound" because Motion is the basis of manifestation and all possible kinds of motions are present in their highest and integrated state in the unmanifest as "Sound" or *Shabda.* It is this mysterious, mystic and incomprehensible "Sound" which is called *Shabda Brahma* and it is from this "Sound" which is inherent in *Brahma Chaitanya* that all possible forms of motion which underlie the phenomena of the manifested universe are derived. It will be seen from what has been said above that *Brahma Chaitanya* and *Shabda Brahma* have practically the same significance as the *Shiva* and *Shakti Tattvas* and must therefore be polar in their nature.

It is because of this polarity that every principle when it descends into the realm of the manifest in its many differentiated forms has always consciousness in the background and all natural forces through which these functions are carried out are therefore conscious forces and not dead or insentient agents of Nature as assumed by Science. It is this association of consciousness or "intelligence" with every natural force and process which accounts for the perfect coordination and adjustment which is always taking place everywhere in nature, a fact which orthodox scientists marvel at and ignore.

Again, because the potentiality for the expression of a function is always present in the unmanifest as *Shabda* or unmanifest "Sound" that a Principle always descends into the realm of the manifest as *Nāda* which in Sanskrit means manifest "sound". Every principle which is a function of consciousness has therefore its counterpart in the world of the unmanifest in *Shabda Brahma* and in the world of manifestation in *Nāda*. *Nāda* is thus the vehicle of consciousness in manifestation and all the different functions of consciousness are thus represented as characteristic vibrations in *Nāda*.

As consciousness descends lower and lower into the realm of manifestation in the form of progressively lower states of mind, *Nāda* descends *pari passu* in different forms of energy which produce the world of forms and bring about interactions between them.

It will be seen therefore that *Shakti* or Divine Power is nothing but Conscious Energy with Consciousness as the rider and Energy as the horse or vehicle for the expression of consciousness. In its highest form this Conscious Energy can hardly be distinguished from the Divine Consciousness or the *Shiva Tattva*. In its lowest form in the physical world it can hardly be distinguished from the apparently insentient natural forces and energies with which Science deals.

The expression of integrated and potential Conscious Energy in the realm of the manifest always takes place through the Point or the *Bindu*. So manifestation is really a dual process. On the one hand integrated consciousness descends lower in the form of differentiated states of the mind, and on the other, integrated "Sound" which is a vehicle and instrument of consciousness becomes differentiated into innumerable forms of energies, in the form of forces which are potential and motions which are kinetic. It is these motions which provide the structure of all forms and these forces which regulate and control the creation, maintenance and destruction of the structures. Manifestation is too vast and complicated a process to enable us to classify or comprehend these differentiated forms of *Nāda* in their entirety, but if we understand these underlying principles in their broad outline, it will help us to understand to some extent the nature of *Nāda* and its relation to *Bindu* and *Kalā*.

Let us now take for consideration the third constituent of the mechanism of manifestation—*Kalā*.

Kalās are the differentiated forms of a principle of manifestation and correspond to the coloured lights of the spectrum when white integrated light is dispersed by a prism. As pointed out in the beginning each principle in its integrated form in the unmanifest breaks up into its own characteristic spectrum of *Kalās* in the manifest. *Kalās*, about the real significance of which there is a great deal of uncertainty and confusion, will thus be seen to be not any determinate things but differentiated and active forms of Divine functions which cannot be determined and defined exactly. Their nature is thus very similar to the nature of coloured lights in a spectrum.

The important point to note now is that the differentiated forms or *Kalās* of every principle or *Tattva* come into existence only when the "Sound" embodying these principles in the unmanifest passes through a *Point* or *Bindu*. *Kalās* are thus the features of the manifested state and cannot be discerned in the unmanifest state, though they must be present in it potentially. They indicate the activity corresponding to each function. In fact, the Sanskrit word *Kalā* is derived, from the root *Kala* which means "to wield", "to put on", "to impel", etc. It implies exercise of a function or engaging in activity appropriate to it. When a function is exercised the corresponding *Kalā* can be considered to come into play.

Tāntrik and other scholars, in order to develop the concept of *Kalā* and give it a scientific appearance, have tried to enumerate and classify *Kalās* of different principles. These enumerations and classifications are arbitrary, not clearly defined and without any rational basis. They vary widely with the mental background of the scholar and his conception regarding the nature of these principles and their functions and generally confuse the average student who is unable to understand clearly what a *Kalā* is essentially. A careful analysis of these *Kalās* will show the student that these classifications merely try to indicate the differentiated functions and corresponding powers of the principles concerned, a task which is wellnigh impossible on account of their complexity and indeterminate nature. It is better, therefore, to be clear in one's mind with regard to the essential nature of *Kalās* and leave the question of their classification open.

It will be seen from what has been said above how the Occult concept regarding the nature of *Nāda, Bindu, Kalā* serves to throw light on the ultimate basis of the mechanism of manifestation and gives us some idea with regard to the nature of the *Krama* which underlies the manifested universe and which has been referred to in the *Yoga-Sūtras* (III-53, IV-32, IV-33). A definite though only broad idea of Cosmogenesis is an essential part of the Occult Doctrine but unless we have a clear idea with regard to the essential nature of the fundamental factors which are involved in the mechanism of manifestation we are likely to miss the inner significance of many processes involved in Cosmogenesis.

In order to understand how the principles existing in their integrated form in the Unmanifest are differentiated into their respective differentiated forms on passing through the Point and thus provide the complex and vast paraphernalia of a manifested system let us take some of these fundamental principles to illustrate this process of differentiation. The prefix *Mahā* in Sanskrit means "great" and when attached to any noun generally indicates the highest or ultimate form or aspect of the thing for which the noun stands. When used in connection with the principles which form the basis of a manifested system it generally indicates the integrated form of the principle as it exists in the unmanifest.

Let us begin with *Mahākāsha*. *Mahākāsha* is the ultimate principle of Space as it exists in the unmanifest. When differentiated by the *Mahābindu* it takes the form of the different mental spaces in which the mental worlds of the different planes function. It is because these spaces are created by the mind, in Divine or individual ideation, that they have no existence in themselves and are therefore called illusory. It is also for this reason that the spaces produced in the realm of manifestation are called *Chidākāsha* or "mental space" in contradistinction to *Mahākāsha* the principle of space in the Unmanifest.

The same remarks are applicable to *Mahākāla* the ultimate principle of Time. According to the Occult Doctrine different measures of time operate on the different planes of existence corresponding to different mental spaces. These different measures of time make it possible for a greater or smaller number of mental images and therefore experiences to be condensed in the same period of time, judged by our standard. As events in the lower worlds of illusion are in a mysterious manner a shadow of what happens in the spiritual worlds (*Yoga-Sūtras* IV-22) it is necessary that the happenings on the different planes of existence, though different, should keep in step. It is the different measures of time which make this possible. These different measures of time may, therefore, be considered as the differentiated forms of *Mahākāla* in the realm of manifestation. They are the *Kalās* of *Mahākāla*.

According to the Occult doctrine each solar system in the universe is the expression of the life and consciousness of a mighty Divine Being who rules and guides its evolution from within by pervading and thus controlling it. That is why the Presiding Deity of a solar system is called an *Īśvara* which means "the Lord and Ruler" in Sanskrit. There are billions of solar systems scattered throughout the universe which is so vast that light takes millions of years in travelling from one end to the other if there is an end. Are these solar systems functioning independently and haphazardly? Yes, according to Science. No, according to Occultism. According to the Occult doctrine, all these innumerable solar systems are the infinite expressions in the realm of manifestation of the Logoic Principle which exists eternally and potentially in the Ever-Unmanifest and which is generally referred to as the Unmanifest Cosmic Logos. Although apparently independent and separated by tremendous distances from each other all the solar systems are in intimate touch on the inner spiritual planes and functioning in perfect harmony and co-ordination because their innermost consciousness is centred in the Cosmic Logos, the Presiding Deity and Ruler of the whole Cosmos. That is why He is referred to as *Maheshvara* (*Mahā Īśvara*). Here again, we may consider these Solar Logoi as the *Kalās* of the Cosmic Logos, the Logoic Principle in the Ever-Unmanifest.

We could take in this manner principle after principle as it exists in its integrated form in the unmanifest and trace its infinite and various expressions in the realm of the manifest, but it is not necessary to do so. The student should have been able to grasp the underlying idea and can himself undertake this kind of exercise to clarify his ideas regarding these fundamental doctrines of Occultism. The student of Occultism should not have towards these doctrines the negative attitude which the followers of orthodox religions have towards their scriptures—to repeat them as a routine without making any serious effort to understand them or to apply the spiritual truths to their lives.

In order to understand the mechanism of manifestation properly we have to grasp not only how the integrated principles in the unmanifest are differentiated in the manifested worlds but also how these differentiated states of the different principles are related to one another. For, on the harmonious relationship of these differentiated principles depends the perfect co-ordination and adjustment of forces and processes we find everywhere in Nature. These differentiated states must not only be related but harmoniously related because they are all derived from one perfectly harmonized Super-integrated State or Principle existing at the heart of the universe.

The manifested universe, as has been pointed out in many contexts, is a repeated projection and expression of this Ultimate Reality at the Centre. In this projection the principles are differentiated further at every step but their mutual relationship at each level remains the same or rather analogous and it is in this manner that the unity and the harmony of the whole Cosmos is maintained in spite of the complexities and apparently unrelated movements which are taking place everywhere. The behaviour of water in a round basin which is being centrifuged from the centre provides a good illustration of the conditions which exist in such a system. The water everywhere is in constant motion and change but the behaviour of particles which are equidistant from the centre is approximately similar and their mutual relations are also approximately the same. And at the heart of this water, in a state of flux, there is the ideal centre in which there is absolutely no movement and no change.

One of the most astonishing features of the progressive experiences which are gained in *Samādhi* when consciousness expands, step by step, and approaches the centre of its origin, is the wonderful unity and harmony which is maintained in the tremendous and sometimes cataclysmic changes which take place in consciousness. The changes are sometimes most fundamental and unexpected but there is no feeling of bewilderment and lack of comprehension at any step. Rather, there is a far deeper and clearer understanding of everything, known or unknown previously. The whole universe seems to be dissolving in a sea of Light which is the very stuff of understanding and perfect comprehension.

I think it would be helpful if we consider specifically the kind of harmony which is maintained between principles that govern the phenomena of the different planes. We have seen that the phenomena of the different planes though vastly different are yet based on some fundamental principles of existence which are an integral part of the mechanism of manifestation. The number and nature of these principles is impossible to determine and know definitely on account of the limitations which are inherent in the intellect and intellectual knowledge. But we can spot at least some of these basic verities of existence such as average size of particles, wavelengths of vibrations, dimensions of space, measures of time etc. It is these features, which characterize the structure of the different planes, between which perfect coordination and harmony are maintained in the process of manifestation, and it is upon the maintenance of this coordination and harmony that the unity of consciousness functioning on the different planes depends. This is a subtle idea which is difficult to grasp but if the student ponders over it deeply he will see that it must be so, if the universe is the expression of one Ultimate Reality which is Perfect, Whole and Integrated.

CHAPTER XIV

THE INTEGRATED STATE—I

Essential Characteristics

As consciousness recedes towards its source in the practice of higher Yoga, all states and expressions of a particular principle are seen to draw together progressively and merge into one another until they are seen ultimately to be derived from a single principle. At the same time all the principles underlying manifestation, in their turn, are seen to draw together and merge into one another and in the ultimate stage are seen to be the differentiated aspects only of One Principle. It is for this reason that this Ultimate Principle or the Absolute is referred to as the Super-integrated State and is the highest and subtlest state in existence.

The conception of the Integrated State which is really an aspect of the conception of the Ultimate Reality is not merely a philosophical question of fascinating interest, not merely a religious question of vital importance to man but also a scientific question of the greatest significance to scientists. Apart from the fact that it throws light on such questions as the origin and nature of the universe—questions which are at least of academic interest to Science—it has great scientific value because it seems to provide some clues for unravelling many riddles of Science and to answer to some extent at least many questions which have been puzzling scientists for sometime.

Questions concerning such fundamental problems as the infinite source of energy found in the universe, the nature of radiation and the electron, the relation of Space and Time, etc., have been bedevilling the scientific world for some time and yet no satisfactory solution seems to be in sight. In fact they have given up the hope of finding solutions to such problems in despair. If they consider carefully the conception of the integrated state and other Occult doctrines related to it, light will perhaps begin to dawn gradually and they might at least begin to realize in which direction they have to proceed in order to find a satisfactory solution.

But it must be stated at the very outset that the concept of the Integrated State is based on the Occult doctrine that the basis of the manifested universe is mental and mind is a differentiated form of Pure Consciousness. So the relation of an Integrated State with the differentiated states and expressions which are derived from it cannot be a mechanical relation which can be expressed neatly and concisely in terms of a mathematical formula. Only what is completely in the realm of manifestation and embodied in form can be conceived and understood and formulated. What is in the realm of consciousness and so beyond manifestation cannot be conceived or formulated. Its expressions can be known and expressed but it itself is indeterminate. It will be seen, therefore, that in the relation of the Integrated State and its expressions, or at a lower level, of life and form, only one element of the duality can be subject to the mechanical laws of the manifested world using the word "mechanical" in its widest and subtlest sense. The other element, by its very nature, is inconceivable, inscrutable and unpredictable and being of the nature of consciousness can be known only in one's consciousness as a direct experience.

In order to understand this relation clearly let us first consider the relation of an archetype and its expressions on the physical plane. We very frequently see an archetype expressing itself in countless forms. The archetype of a rose, for example, can find expression in countless roses, all different from one another. It is like the abstract idea of a triangle which can be expressed in innumerable triangles of all shapes and sizes. The relation of an integrated state and its expressions is similar to this but the similarity is only partial and there is an underlying fundamental difference. This difference lies in the fact that an archetype or an abstract idea cannot by itself produce a rose or a triangle. The archetype or the abstract idea of a thing is just like a model in the mental world which provides the essential features of the form but it cannot provide the impulse and the energy to create just as a plan of a building cannot produce a building or a set of blue-prints cannot produce an aeroplane. Something more is needed and this is energy, life, intelligence, all derived from consciousness.

The above fact is of great importance if we want to understand the Occult point of view regarding the nature and origin of the universe in contradistinction from the purely materialistic scientific point of view which regards the universe as a product of blind forces of nature. We should understand clearly that the mind—higher and lower—provides only the substance, the models and the instruments for the production of forms. But the Reality which is behind the mind and from which the mind is derived provides the energy, purpose, design, power of perception—all related to consciousness—in the manifestation of the Cosmos. The equation of the integrated state on the one hand and its expressions on the other can therefore be never completed as symbolized below:

$$F(X)? = x_1 + x_2 + x_3 \; \ldots\ldots \longrightarrow \infty$$

Let us look at the same problem from a different point of view to clarify further our ideas on the subject. It has been pointed out in different contexts that manifestation is a repeated process of projection, involvement and expression of the One and the Ultimate Reality in progressively denser states of matter and less subtle states of mind. This Reality projects itself into a lower state of Itself, is involved in that state by pervading it and then expresses itself through that state in more and more complex and evolved forms. This whole idea has been put in a beautiful manner but in a somewhat different way in the following line from the Hindu scriptures

स्वे महिम्नि स्वयं स्थित्वा स्वयमेव प्रकाशते ।

sve mahimni svayam sthitvā svayameva prakāshate

"He establishes Himself in His own glory and then shines through it in His Divine splendour."

The joint process of projection, involvement and expression referred to above is repeated producing different levels of manifestation which are characteristic of the complex Cosmos that lies hidden within the outermost physical universe with which alone scientists are familiar. It will be seen, therefore, that not only is the Cosmos in its total aspect an expression of the Ultimate Reality but that this Reality is involved in manifestation in the most intimate manner at every level and in every sphere of existence. And there are two factors present in every aspect of this involvement, one subjective and the other objective. It is these two inseparable polar realities underlying manifestation everywhere which are referred to as Spirit and Matter or as *Purusha* and *Prakriti* and represented symbolically by the interlocked double triangle with which most students of Occultism are familiar.

Before we deal with the nature and relation of the integrated and differentiated states let us first clarify further our ideas with regard to Spirit and Matter or *Purusha* and *Prakriti* which we find inter-twined and inter-acting in every sphere and level of manifestation and producing the infinite variety of phenomena in the Cosmos.

Let us take first the component of this duality which is referred to as Matter. The student should note carefully that by the word "Matter" we do not, in Occultism, mean "matter" in the ordinary sense in which the common man or even the scientist uses the word. In Science the word "matter" is used to denote that part of the manifested existence which is tangible, has "mass" and is corpuscular at its base. This scientific conception of matter has become modified fundamentally by the recent discoveries in physics and chemistry which show that matter is essentially radiation, "bottled-up radiation" as James Jeans has called it. As this is a very vague and nebulous conception the scientist does not now really know what matter really is, or how to define it. The nature of "mass" by which matter is defined still baffles the scientist to the same degree as before, perhaps much more.

In Occultism, we use the word "Matter" in the sense of an ultimate Principle which is the opposite of Spirit or Consciousness and which merges with Spirit or Consciousness to produce the Void state of the Absolute or the Ultimate Reality. Matter is thus the negation or antithesis of Spirit or all that Spirit implies. The word is thus used in Occultism in a very wide and subtle sense which has hardly anything in common with the conception of matter in Science. In its highest conception of " Matter " in Hindu philosophy *Prakriti* or Nature is considered to be *Mahāmāyā* the Great Illusion which obscures the full awareness of Reality and thus makes manifestation possible.

Just as Matter is a Principle so the Spirit also is a Principle but with a fundamental difference. It is the positive element in the duality of Light and Darkness. It is of the essence of Consciousness and so infinite, unpredictable and immeasurable in its nature and potentialities.

It has been pointed out already that manifestation is a repeated projection, involvement and expression of the One Reality in progressively denser states of matter and subtler states of mind. In the different levels of manifestation which result from the repeated process the higher state is found reflected in the lower state in a mysterious manner at every stage of descent into a lower state. The higher state is not only reflected in the lower state *as a whole* but it is also differentiated to a greater extent, thus leading to the appearance of more numerous and diverse states and principles. It is for this reason that at every level of manifestation, every principle is found present in modified and diverse forms though it may be difficult to trace it in the ramifications of its countless expressions.

The reflection of the higher state into a lower state *as a whole* means really the reflection of the state in an integrated form. For the chief characteristic of an integrated state as distinguished from a differentiated state is the sameness and wholeness of the state in every part, although there are really no parts or aspects in such a state. It would perhaps be more appropriate to say that it is not possible to distinguish one aspect from another aspect in an integrated state although all its aspects are present in it in a potential form and can manifest as required when the required conditions are present. It is in this sense that the integrated state is called *Nirviṣhesha* " without any particular features ".

An idea of the distinction between an integrated and the differentiated states derived from it may be obtained to some extent by comparing white light with the coloured spectrum derived from it. White light is integrated, whole and the same in every part. The coloured spectrum is differentiated, broken up into lights of different colours, the colour and frequency of light in every part differing from the colour and frequency of light in any other part.

Although the integrated state is a transcendental state and above the differentiated states of the mind in which we live, still, it is necessary for the aspirant to strain his intellect to the utmost and try to understand it intellectually as far as he can, with his present limitations. For, in the nature of the integrated state are hidden not only all the secrets and solutions of enigmas of Philosophy and Science but also the supreme secret of his own existence which he is seeking and which will set him free from the illusions and limitations of life on the lower planes.

This state can be " known " in the real sense only by the Yogic technique of " knowing by becoming " i.e. by suppressing all the modifications of the mind at the lower level when the unmodified state of the mind emerges from within the mind and can be experienced. This is a state of pure consciousness from which the modifications of the mind were derived and by which these modifications were illuminated. It is this consciousness as distinguished from the modified mind which has all the features of the integrated state in a greater or lesser degree. " In a greater or lesser degree " because the Integrated State like everything else in the realm of manifestation is governed by relativity. It is the repeated reflection of Reality in lower states of Itself which results in the appearance of these relative integrated states at different levels. This question has been discussed in some detail in dealing with the nature of *Asamprajnāta Samādhi* and the relation of " Mind and Consciousness ".

But before we pass on to discuss the nature of the integrated state it is necessary for the student of Occultism to realize that this state is not merely one of the hypothetical assumptions or theoretical abstractions in which the literature of academic philosophy or theoretical science abounds. It is a real state which can be experienced and known definitely, more definitely and realistically than any other thing in the phenomenal world of the mind can be experienced. In fact, the direct experiencing of all these relative integrated states, one after another, and ultimately of the true integrated state, is the real object of Yoga. And it is from these experiences that all knowledge concerning the transcendental truths of existence is derived. And if Science wants to " know " the ultimate secrets of Nature and not flounder eternally in uncertainties and mathematical abstractions, it will have to leave its present methods based on external physical apparatus and take to the time-honoured and tested path of Occultism based on Yogic discipline.

It has been pointed out already that the nature of the integrated state is a question which is of great interest and significance in all the three fundamental fields of knowledge —Religion, Philosophy and Science. It is of importance in the sphere of religion because it throws light on the nature of

Divinity or the Reality which lies hidden at the basis of the manifested universe. In fact, our conceptions about this state are derived mostly from the conceptions of Divinity found in the scriptures of the great religions of the world or the conception of Reality found in Occultism. The conception of this state is of importance in philosophy because it throws light on the nature and the origin of the universe, on the nature of manifestation and the relation of the Manifest and the Unmanifest. It is of importance in science because it throws light on many problems which have arisen during the recent developments in many fields of science, developments which have knocked out the bottom of the whole structure of scientific theory and philosophy. The nature of infinity and its relative nature, the origin of dualities which are found hidden at the basis of all natural phenomena, the infinite source of energy which is apparently hidden somewhere, the alternation of opposite states and the constant tendency found in Nature to revert to a state of equilibrium, all these and many more problems which bedevil the scientific world and have driven scientists to the verge of despair cannot be understood even partially except in the light of the philosophical conception of the integrated state.

In trying to gain some comprehension of the integrated state it will help us if we start with the inner meaning and significance of some words used in Sanskrit in reference to the Reality underlying the universe. The two words which are used very frequently in this connection are *Nirvishesha* and *Aparināmita*.

Nirvishesha means literally " without any particulars (but containing all particulars in a potential form)". These " particulars " will depend upon the nature of the integrated state and become apparent when the integrated state becomes differentiated. Let us take a simple integrated state like white light. White light does not contain any particular colour as such but still it contains all colours potentially. All these different constituent colours become manifest and can be seen together when white light is dispersed by means of a prism in the form of a spectrum. They become manifest singly when white light falls on objects of different colours which absorb part of the white light and reflect back the remainder.

White light represents a simple integrated state on the physical plane. Let us now take a far more complex integrated state, the human consciousness. Human consciousness even though it is the lowest expression of the Divine Consciousness, is one of the most mysterious and wonderful things in the world, although in our preoccupation with worldly affairs, and owing to our lack of sensitiveness, we are quite unaware of this fact. Owing to the lack of sense of discrimination and values we are straining all our faculties to find out what is present on the moon or even on the distant stars, but here is our human consciousness containing hidden within its deeper recesses infinite knowledge, power, bliss and all the secrets of Nature and we do not care for this fact.

The wonderful nature of human consciousness and the fact that it contains within itself potentially an infinite number of attributes, functions, faculties, etc., which can manifest according to the needs and circumstances present, can be seen in our ordinary experiences. Let us imagine a king sitting in meditation. In this state of his consciousness all human functions and attributes are in abeyance and present only in a potential form. He gets up and mixes with the members of his family. He immediately becomes and functions as a father towards his children, as a husband towards his wife, as a master towards his servants. All these different functions are present potentially in his consciousness and spring forth into action as and when required without any conscious effort on his part.

Take another aspect of his integrated state of consciousness which finds expression in different attitudes and emotions. He behaves towards members of his family with love and affection, towards his subjects with sympathy and consideration for their welfare, towards those who are in conflict with justice, towards criminals with sternness and strict conformity with law. All these different attitudes and emotions are present in his consciousness potentially and manifest as and when required.

Take another aspect of the integrated state of his consciousness, that which is related to his mental faculties. The king tries one of his subjects for a crime. A proper judgment comes out of his mind as needed. A new palace has to be constructed. A blueprint for the building appears in accordance with his wishes. An enemy attacks his kingdom. A strategy of defence is formulated and put in action at once.

So his consciousness is capable of bringing out of itself any function, any attribute and faculty as and when needed although no indication of the presence of any of these is present when the king is sitting in meditation. It is true that the functions, attributes and faculties etc., have to be unfolded gradually by the process of evolution but the fact remains that they are all present together in their integrated form in his consciousness and can come out as needed in his life.

All this is true in the case of ordinary human consciousness which is comparatively very little unfolded. Imagine the potentialities which must be hidden in the integrated state of Divine Consciousness which can bring out of itself any attribute, any power, any function needed in running the vast machinery of a manifested system. And try to imagine the nature of the experience which is gained in the higher stages of Yoga when the Yogi is able to get a glimpse into this Divine Consciousness, integrated or differentiated.

Aparināmita means "changeless". What is *Nirvishesha* or "without particulars" must also be changeless. The reason for this lies in the very nature of the integrated and differentiated states. It is only when the differentiated state appears, that there can be particular states which can be distinguished, one from the other. And it is only when there are particular states or aspects or parts which can be distinguished, one from the other, that there can be any change. For, change is nothing but the transposition of different parts of a whole or replacement of one state or aspect by another within a whole. And where there are no parts, no states, no aspects which are different and can be distinguished from one another there can obviously be no change. So change is a feature of the differentiated state and the integrated state must be "changeless". And since mind is a differentiated product of consciousness it alone must be subject to change while consciousness must be above change.

The two attributes of integrated consciousness discussed above show clearly how this state of consciousness must also be above time and space. For, how do we perceive or become aware of space? By the presence of different things which occupy different parts of the whole and the spatial relationships which exist between these things. If there are no parts in the whole there will naturally be no spatial relationships and so no means of perceiving space. In other words, the integrated state of consciousness will be above space.

The same considerations hold in the case of time. We become aware of the passage of time as a result of the changes which we notice in our environment, and mind, or in a deeper sense, in the different parts or states of the whole which are present side by side. If there are no parts, no states which can be distinguished from one another there can be no change, and if there is no change, the awareness of time will be absent. In other words, the integrated state of consciousness will be above time.

It will be seen therefore how the conception embodied in the Sanskrit word *Nirvishesha* provides a clue to the other attributes of the integrated state of consciousness. It is obvious that it is differentiation and the consequent appearance of different parts, states, aspects, etc., which lies at the basis of the phenomenal world and creates the illusions of change, time and space. The integrated state of consciousness or Reality must therefore be above these illusions but in a relative sense. It is only the Super-integrated State of the Absolute which is above these illusions in the absolute sense.

CHAPTER XV

THE INTEGRATED STATE—II

Infinite Expressions Of Consciousness

Having considered the nature of the integrated state in a general way let us now see how this conception throws light on the problems of Philosophy and Science, problems which seem to defy solution and have created a great deal of confusion in scientific thought. It is because Science is on very sure ground as far as facts and laws of the purely physical world are concerned and can thus utilize with complete confidence and exactness the forces of Nature in investigating the phenomena of the physical world, and especially in producing spectacular results, that the confusions and contradictions of scientific conclusions are tolerated and even accepted blindly. But for this fact, the general theories of Science concerning life in general and particularly its adventures in the realm of philosophy would not have been taken very seriously even if not rejected outright.

The human mind is prone to worship success blindly and if any individual or group of individuals have achieved remarkable success in any line they are immediately put on a high pedestal and thought fit to pronounce judgments on any and every question, even though they have not the slightest qualification or competence for doing so. The opinions of scientists with regard to the origin, nature and destiny of the universe, their rejection of the highest truths concerning human life and the universe which have been proclaimed and emphasized again and again by the Great Teachers and Prophets of world religions, their refusal to take serious notice of the doctrines of Occultism, all these things should be considered in the light of what has been said above.

Occultism does not claim to offer a cut-and-dried solution of these insoluble problems. It regards the deeper problems concerning man, God and the universe as too stupendous to be solved or even comprehended by the puny human intellect. But it claims that the universe is a Cosmos created and guided by a Supreme Intelligence and not a madhouse self-created by "a fortuitous concourse of atoms" and interaction of blind forces of Nature, that there is a Divine Reality underlying and pervading it and there is therefore harmony, beauty, purpose hidden within the apparent conflict, confusion and purposelessness which we see outside owing to our limited vision, that it is possible to know directly this Reality and the inner worlds of beauty and harmony and power which support and guide the stupendous physical universe, that all this knowledge has been gained by well-tested and tried methods of Occultism, that it can be gained again and verified by any individual who makes the necessary effort. But it can be gained only by the time-honoured methods of Occultism and not by methods demanded and dictated by scientists who know nothing about these things and care less.

Let us now proceed to examine in a general way some of these problems of philosophy and see whether Occultism helps us at least to understand them intelligently instead of confusing and confounding us as modern theories of Science do. Let us see first how the conception of the integrated state throws light on some of these problems.

We have seen in the last chapter that the integrated state is whole, changeless, the same everywhere and contains no parts or states which can be distinguished from one another. Obviously, such a state will be above the jurisdiction of mathematics. For, in mathematics we deal with relations of things with one another, and if there are no things or parts within the whole, the question of relationships existing between them does not arise. And so such a state should be above the laws of mathematics in itself. But as has been shown in the previous chapter an integrated state contains within itself in a harmonized and potential form all those things which can be derived from it on differentiation. It must also therefore contain within itself in a harmonized and potential form all those laws of mathematics which regulate the relations of things which appear from it when differentiation and manifestation takes place. This is an extremely fascinating though subtle idea which the student should try to grasp. If there are two states, one integrated and the other differentiated, and the differentiated state can merge in, or appear from, the integrated state then the integrated state must contain in an integrated form not only the differentiated constituents but also their mutual relationships which determine the relations of the constituents among themselves when they appear from the integrated state.

A concrete example taken from the phenomenon of light will make this point clear. White light contains within itself not only the coloured light of the spectrum but it must also contain within itself the mathematical relationships existing between these coloured lights when they are dispersed by a prism. That is how, when the coloured lights appear in the form of a spectrum, they are arranged in a particular order and particular frequencies of vibrations which can be calculated with the greatest possible exactitude. That is also how, when different phenomena of light take place under exactly the same conditions, they are reproduced exactly in the same form. In fact, the whole philosophy and technique of scientific investigation is based upon this underlying principle—a phenomenon is produced exactly in the same form when the conditions are exactly similar.

The same principles hold true when manifestation takes place from an unmanifest state and differentiated states appear from the integrated state on a macrocosmic scale. According to the Occult doctrine the whole of the manifested universe becomes *laya* or merged in and exists in the Consciousness of Śiva when *pralaya* takes place and it appears again from His Consciousness when *srishti* or manifestation takes place as expressed beautifully in the following Sanskrit text:

शेते तिष्ठति सर्वं जगत् यस्मिन् सः शिवः शम्भुः विकाररहितः ।

Shete tishṭhati sarvamjagat yasmin saḥ Shivaḥ Shambhuḥ vikāra-rahitaḥ

"Śiva or Shambhu is that Supreme flawless Reality in whose consciousness the universe exists in a state of quiescence during the period of *Pralaya*."

We can see easily that when the universe becomes *laya* or merged in the integrated consciousness of *Siva* or in the unmanifest all the laws of Nature which regulate the activities of the phenomenal world during a period of manifestation, and the mathematical relations which form the basis of these natural laws, must also merge in and exist in a potential form in the unmanifest, and must emerge from the unmanifest again to regulate these activities on manifestation.

So, while an integrated state in itself is above the laws of mathematics it not only provides all the differentiated states on manifestation but also regulates their mutual relations and interactions according to mathematical laws. This principle is the mathematical equivalent of the principle enunciated before, according to which an integrated state contains no differentiated states as such but is the potential container of all such states. Without the existence of this principle periodical manifestation could not invariably take place in a predetermined, ordered and perfect form and all its activities could not be regulated by natural laws which work with mathematical exactitude. And also, without this principle the Reality which underlies the universe and which is generally referred to as God could not be a free, Self-determined and Self-contained Divine Principle but would be like the reality which Science has found as the basis of the physical universe—a system of blind and unconscious abstractions and laws.

Before we pass on to discuss the implications of the above mentioned mathematical relation existing between an integrated state and the differentiated states derived from it, we may ask in what form the laws of mathematics exist in the integrated state and how they may be represented when present in this state. To answer this question we may recall what was said about the symbolic representation, in mathematical terms, of the Absolute which is another name of the Super-integrated State. We saw that this State can be represented only by zero not only because it is a Void in one of its aspects (zero is called *Shūnya* in Sanskrit and *Shūnya* means Void) but also because it has limitless potentialities. Just as we can derive from zero any quantity as a pair of equals and opposites so the whole manifested universe has been derived from the Absolute as a pair of Opposite Principles called *Siva* and *Sakti Tattvas* (by Primary Differentiation) and from these two fundamental and transcendental Realities every reality in the manifested universe is derived as indicated in the following Occult maxim:

Shiva-Shaktyātmakam Vishvam

But there is a further Differentiation of these two polar and opposite Principles, each Principle giving rise to a triplicity as shown below:—

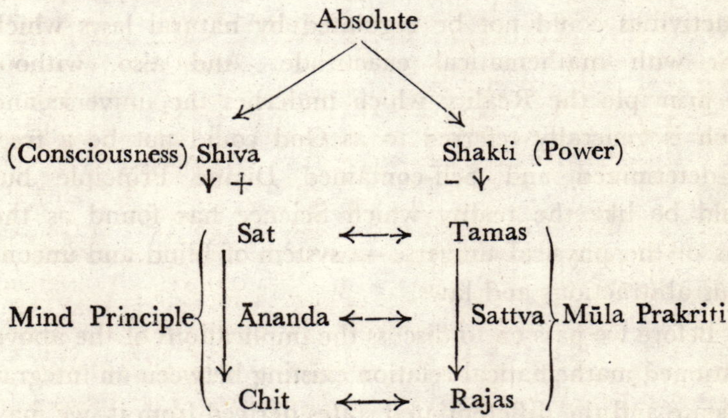

It is this Mind Principle and *Mūla Prakriti* derived from Consciousness and Power respectively which together constitute the Logoic Principle and are the source of the infinite number of universes which continue to appear from the Ever-Unmanifest in the eternal alternation of *Srishṭi* and *Pralaya*. Both, the Mind Principle and *Mūla Prakriti*, have infinite potentialities as can be seen easily by examining the Occult conception of the nature of the Unmanifest and the Manifest and their mutual relation. Firstly, an *infinite*

number of universes continue to appear from the Logoic Principle in the eternal alternation of *Srishṭi* and *Pralaya*. Secondly, the number of Logoic manifestations in the form of galaxies and solar systems is infinite. Science, with its powerful telescopes has been able to discover billions upon billions of galaxies and solar systems and has not found the limit of even the physical universe. Thirdly, each manifested system is infinite not only in its potentialities but also in its expressions. In view of these infinite potentialities in the Ever-Unmanifest which are the source of infinite expressions in the manifest, the unmanifest or the integrated state can be best represented by the two most mysterious, potential and indeterminate entities of the mathematical world, O and ∞. The zero symbolizes the potentiality of the integrated state because *anything* can appear from it in equal and opposite forms. Infinity symbolizes the potentiality of the integrated state because anything which appears can be *infinite* as regards number, quantity, intensity or forms of expression.

It is important to note that both zero and infinity are indeterminate and undefinable in mathematics and thus reflect and symbolize the integrated state in a natural manner. They are both the potential containers of infinite expressions and yet containing no such expressions within themselves. We should also note carefully that it is zero which first gives rise to equal and opposite positive and negative quantities and these quantities can then assume infinite forms provided they remain balanced and add up to zero. In the same way an integrated state first gives rise to two equal and opposite differentiated states and then each of these opposite states can assume infinite number of forms or expressions, the totality of both remaining the same and thus maintaining the balance of the Void state in the Absolute. This fact has now been corroborated by recent scientific developments and has inclined the scientists to suspect the presence of negative worlds which balance the positive worlds which we know.

The fact that an integrated state, in itself, cannot be governed by the laws of mathematics is of great significance and provides a clue to a partial solution of many problems of Philosophy and Science. For, if a state is inherently above the jurisdiction of mathematics all the yardsticks which are the paraphernalia of scientific measurements such as number, quantity, intensity cannot apply to that state. It must be beyond the restrictions and limitations of all ordinary natural laws and instruments of measurements. In other words this state must be free from the ordinary limitations of manifested existence.

This is exactly the conception of all great religions concerning that Reality which is referred to vaguely as God. God is considered to be of the essence of consciousness and to express Himself as consciousness in the manifested world. The Divine Consciousness is considered to be free from the limitations which bind matter and spiritual entities imprisoned in material vehicles.

The conception of God in Occultism is more definite, complex and richer but the essential nature of this Reality is best indicated by the philosophical conception of the integrated state which has been discussed above. In Hinduism also this Reality is called *Nirvishesha*, *Akhanda*, *Pūrṇa* etc., words which merely indicate different aspects of the integrated state. It is because this Reality is above the laws of mathematics and not restricted by limitations of quantity, quality, number, intensity, etc., that it is relatively free and Self-determined and can exercise the functions of an *Iśvara* or Logos. But it must be remembered that this is true only of the Reality itself in its unmanifest and integrated states. What comes out of that Reality or, in other words, is created, is in the realm of manifestation and differentiated states and therefore subject to all the limitations of such states, and governed by mathematical laws.

It is, however, necessary to bear in mind the relativity of integrated states, even though it is extremely difficult to comprehend what relativity of integrated states means. All integrated states below that of the Absolute must be relative. It is only in the Absolute that all integrated states find their culmination and harmonization. That is why it is called the Super-Integrated State. These lower integrated states may therefore be considered as the differentiated states of the Super-Integrated State.

It is also necessary to distinguish between these integrated states derived from Consciousness and those derived from Power or from *Purusha* and *Prakriti*. It is only those derived from Consciousness which have the attributes of Consciousness or *Purusha* such as freedom, self-determination etc. Those derived from Power or *Prakriti* are within the realm of manifestation and subject to the laws of Nature and mathematics. But the capacity for infinite expressions is common to both. Not only can each centre of integrated consciousness express itself through an infinite number of centres but an archetype can also express itself through an infinite number of forms. This is necessary because the objective aspect of Reality or *Prakriti* has to provide for the infinite expressions of the subjective Reality.

Let us consider these different aspects of the integrated state of consciousness in relation to Logoic manifestation, for, it is in the realm of the macrocosm that they appear in their most multifarious and prominent forms. But since the human soul is also a microcosm in a very partial state of unfoldment the same aspects are present in the expressions of human consciousness also, although in a less prominent form.

Since the Divine Consciousness of a Logos is the basis of a manifested system and is the source of all things needed in that system it must contain within itself every thing which is needed in such a vast and multifarious field of evolution. Even a cursory examination of this vast field will give us some idea of the tremendous potentialities and resources of Divine Consciousness. But all these things which may be needed in the manifested system should be present in the Divine Consciousness in an integrated form and not as differentiated states because it is only such an integrated state which can supply *all* possible differentiated states which are based upon and derived from it. This is inherent in the very relation of an integrated state and corresponding differentiated states.

It is necessary to remember that such an integrated state can contain within itself many sub-integrated states, each constituent integrated state being related to its corresponding differentiated states. It is thus really a super-integrated state though this phrase is used generally only for the Ultimate Reality or the Absolute. Even in the field of physical phenomena we can see somewhat different kinds of integrated states belonging to the same category being present in a kind of superintegrated state. Thus the radiation coming from the sun is an integrated state containing within itself not only the visible spectrum of coloured lights but also vast ranges of infra-red and ultra-violet radiations.

The consciousness of the Logos of a manifest system is such a super-integrated state containing within itself many kinds of integrated states. We may divide these constituent integrated states into two broad sub-divisions—the integrated state of Divine Functions and the integrated state of Divine Powers. In Hinduism this integrated state of Divine Functions is referred to as *Mahā Deva* and the integrated state of Divine Powers is referred to as *Mahā Devī*. The Sanskrit prefix *Mahā*, as has been pointed out in other contexts, always denotes the ultimate and subtlest state of any reality in existence which can exist in innumerable forms and at different levels of manifestation. This ultimate state is nothing but the integrated state from which all differentiated states of that particular category are derived. All such names containing the prefix *Mahā* e.g. *Mahākāla*, *Mahākāsha*, *Mahāmāyā* should be considered in this light if we are to understand their real significance and the nature of the realities indicated by them.

It is these differentiated Divine functions and powers which are derived from *Mahā Deva* and *Mahā Devī* which are symbolized in Hinduism in the forms of various *Devatās* and *Devīs*. Corresponding to every Divine function there is a Divine power which enables that function to be exercised. This power may be considered to be the negative aspect of the corresponding positive function and that is why the *Devī* is referred to as the consort of the corresponding *Devatā*. It should not be considered as separate and independent from consciousness but as an inherent attribute of consciousness. The difference in the symbolic form representing a Divine function and its corresponding power is due to the difference in the expressions of the two. It is because Divine Consciousness is such an integrated all-pervading state that in a manifested system any function can emerge from this state, as and when and wherever required, and also the corresponding power which enables that function to be carried out.

Although it is impossible to enumerate or classify the Divine functions and powers some idea of these functions and powers may be obtained from the nature of *Ashṭa Mahā Siddhis* referred to in Yogic literature. This ultimate *Siddhi* is eight-fold, its eight different constituents being referred to in Sanskrit as *Aṇimā*, *Mahimā*, *Laghimā*, *Prāpti*, *Prākāmya*, *Īshattva*, *Vashittva*, and *Kāmāvasāyattva*. The use of the prefix *Mahā* in the name of this *Siddhi* is significant not only because it is the highest *Siddhi* attainable but also because it is an integrated form of the highest *Siddhis* needed in the exercise of Divine functions. It is the integrated Power of integrated Consciousness and its constituent powers can be exercised only by the *Mahātmā* who is established in the realm of integrated consciousness. That is also why these *Siddhis* are referred to as *Siddhis* of *Shiva*, the symbol of integrated Divine Consciousness.

The Consciousness of the Logos is the source not only of countless Divine functions but also of the vast and continuous series of *mental* phenomena at different levels which a manifested system really is. In the same way His Power is the source not only of countless expressions of Divine powers but also of the innumerable *mechanisms* of different complexity which Divine Power as *Prakriti* provides as the basis of the mental phenomena. It is true that both types of expressions take place not only through the Logic Centre (*Mahābindu*) but also through the Monadic Centres (Bindus) which are present in a manifested system, but the background of all of them is the integrated Divine Consciousness.

We may also dwell for a while on the kind of changes which take place when consciousness descends from one plane of manifestation to another. Again, the analogous behaviour of light will enable us to understand to some extent the changes which are brought about. When white light is passed through a transparent coloured medium it is deprived of certain constituent colours and the emergent light differs from the incident light in having a lesser number of constituent colours. But the emergent light still remains integrated as has been explained in another context in dealing with the relation of mind and consciousness.

In the same way, one state of integrated consciousness may differ from another such state by being deprived of certain aspects in the descent to a lower plane owing to the limitations which are inherent in the very nature of the lower plane. This idea will perhaps help us to understand the mystery of Divine manifestation on the three Divine planes. According to the Occult doctrine, on the *Ādi* plane, the highest plane of purely Divine manifestation all the three aspects of Divinity are present in perfect harmonization, and function as *Maheśa* or the First Logos. On the next lower plane, the *Anupādaka* plane, one of the aspects, namely, the *Sat* aspect is partially suppressed, leaving the other two to function fully as *Vishṇu*, the Second Logos. On the lowest plane of Divine manifestation, the higher *Ātmic* plane, another aspect of the triple Divine Consciousness the *Ānanda* aspect is partially suppressed, leaving only the *Chit* aspect or the Mind Principle to function as *Brahmā*, the Third Logos.

It should be noted that the suppression of these aspects in the descent to the lower planes is not complete and the Divine Consciousness does not function on the three Divine planes in water-tight compartments. Behind the consciousness of *Brahmā*, the Creator, is present the consciousness of *Vishṇu*, the Preserver and in the background of the consciousness of *Vishṇu* is present the full consciousness of *Maheśa*, the Presiding Deity of the system. It is the same Logos or *Īshvara* who is present on the three planes but exercising different functions owing to the limitations and peculiarities of the three planes. We may therefore imagine the Logos, if we like, as a Triple Reality with the *Chit* or Mind aspect as the outermost, with the *Sat* or Will aspect as the innermost and *Ānanda* or Bliss aspect in the middle, linking the *Sat* and *Chit* aspects. But it is the same Logos who is functioning on the three planes as Ruler, Preserver and Creator. Since the *Jīvātmā* or the Individuality of man is a microcosm reflecting the nature of the macrocosm we see the same mutual relation existing between its three constituent principles—*Ātmā*, *Buddhi*, *Manas*.

CHAPTER XVI

THE INTEGRATED STATE—III

Co-Existence of Integrated and Differentiated States

We have discussed the nature and the mutual relation of the integrated and differentiated states in different contexts in order to understand as clearly as possible the different aspects of the problem. This is perhaps the greatest mystery of spiritual life and the most baffling problem of philosophy. This does not mean that it is merely an interesting academic question upon which we can speculate endlessly to satisfy our philosophical curiosity. It is the most significant fact of Yogic psychology and upon the solution of this problem is based the technique of *Samādhi.*

It has been pointed out in different contexts that pure consciousness being an integrated state not only contains all the differentiated states of the mind in a potential form but contains them in a form which is very mysterious and difficult to understand. It is only when he differentiated states of the mind are completely suppressed in *Chitta-Vritti-nirodha* that the Yogi can become aware of the integrated consciousness which provides the background and the illuminating power behind the modifications of the mind. It is only then that the Yogi can realize the nature of integrated consciousness and its relation with the differentiated states of the mind.

The nature of the integrated and differentiated states and their mutual relation is a problem of the utmost significance in philosophy because upon it depends our conception of the Ultimate Reality which underlies the universe and which is the background of consciousness in the case of every human being. The whole of the manifested universe in all its aspects and degrees of subtlety is nothing but an interplay of the differentiated mental states derived from integrated consciousness functioning at different levels. The manifested universe, as has been pointed out in other contexts, is the result of the repeated projection, involvement and expression of the One Reality in successive stages. At each of these stages the relation of an integrated state with its corresponding differentiated states is involved.

Even in the field of Science the relation of the integrated and differentiated states is the core of most theoretical problems which baffle the scientists. Take the ordinary and familiar phenomenon of light. Hundreds of scientists have been working upon this problem for almost two centuries and trying to unravel its mystery. But the deeper they go into it the more it baffles them. They know of course, in the minutest detail the physical laws underlying this phenomenon and can therefore control and manipulate this form of energy in the most wonderful manner. But when it comes to the fundamental question: what is light?, they find a dead wall facing them. For, light is an integrated state and the nature of an integrated state can be known only when the modifications of the mind produced by the differentiated states are suppressed.

In view of the importance of understanding the nature of the integrated and differentiated states and their mutual relation let us consider some other aspects of these states.

The first point we may note with regard to the integrated state is that it contains the differentiated state within itself in a mysterious form. It must contain the differentiated state within it in some form because the differentiated state comes out of the integrated state. That we cannot get something out of nothing is a self-evident truth. The emergence of the differentiated states from the integrated state under certain conditions therefore shows positively that these states exist within the integrated state even though we cannot discern them in the integrated state. The presence of the differentiated states in the integrated state is analogous to the presence of a photograph in a photographic plate which has been exposed but not yet developed. We cannot discern any change in the undeveloped plate and yet the whole of the photograph is present in it in a potential form.

This fact lies at the basis of the Occult doctrine that nothing is ever really created in the usual sense. It is all a question of the emergence or re-emergence of what already is present in the realm of the unmanifest into the realm of the manifest. For example, when the universe appears into the realm of the manifest after a period of *Pralaya* it is not "created". It has merely become manifest after remaining in a state of quiescence in the unmanifest. Where does it remain in the state of *Pralaya*? In the consciousness of *Śiva* as pointed out in the previous chapter.

It will be seen from what has been said above that if the Ultimate Reality is the ultimate source of everything in the realm of the unmanifest or the manifest it must contain everything within Itself in some mysterious and incomprehensible form even though nothing can be discerned as such in this Super-integrated State. Otherwise, how could everything come out of that Ultimate Reality in the eternal alternation of *Srishṭi* and *Pralaya*.

This concept throws new light on the nature of this Reality which we generally consider quite free from any trace of the differentiated states. In one aspect it is a Super-integrated State which is quite free from differentiated states. In another aspect it is the potential container of all differentiated states, nay, all possible states and therefore not quite free from the differentiated states.

The fact that any differentiated state can come out of the corresponding integrated state even though the integrated state is *Nirvishesha* and not a trace of any differentiated state can be discerned within it also throws light on the question whether the *Purusha* is an actor in the drama of which he is the centre. This enigma has bedevilled Sānkhyan philosophers from the very beginning and made them come to the strange conclusion that the *Purusha* is a mere spectator of this drama and has no hand in its direction or performance. It has been shown in another context that this view is lop-sided and self-contradictory and untenable in view of the Occult doctrine concerning the nature of manifestation and the manner in which the activities of a manifested universe are planned and executed.

The Sānkhyan view concerning the passive role of the *Purusha* in the drama of the universe is not without some justification because the *Purusha appears* to have no active role in this drama and things just happen around him and for him without his taking any part in them. According to Sānkhyan philosophy all this activity and planning is due to *Prakriti* which not only plans all activities for the evolution of the *Purusha* but also provides all the paraphernalia for these activities. How an insentient principle like *Prakriti* can plan and execute the activities of the vast and complex universe, and if *Prakriti* is a conscious Principle, where is its consciousness derived from, all these questions are left unanswered.

The clue to this enigma is provided by the Occult doctrine concerning the nature of the integrated state. We have seen above that every differentiated state or function required in the realm of manifestation can appear from the integrated consciousness even though no differentiated state or function can be discerned in the integrated state. The integrated state can play this dual role because it functions *as a whole* in providing the differentiated states. A particular function does not come out of any particular part of the integrated state because there are no parts or separate states or functions present in this state. The integrated state can provide a *vishesha* or particular function or state as a whole or as a *nirvishesha* principle.

The reason why the *Purusha* seems to have no active role in the drama of which he is the centre is similar. No particular desire or impulse to action can be detected in him and yet every impulse to action can come out of him because his consciousness is an integrated state and acts *as a whole*. This is a rather subtle idea which the student should try to understand carefully. We seem to have all kinds of desires which motivate our actions because we are functioning in the realm of mind. A particular desire and a particular action motivated by it can be isolated and separated from other desires and actions and that is why we appear to be actors in the happenings which take place round about us and appear to be initiated by us. But if our consciousness were to act as a whole and not from a particular desire or particular impulse of will we would not appear to be the actor in a particular situation. The particular thing will appear to "happen" and not to be "done" by us. An ordinary action is a particularized activity and that is why its doer can be isolated and identified. But if it is not a particularized activity having its source in a particular part of our mind but the activity of integrated consciousness which functions *as a whole* it will not be possible to isolate and identify the actor. And yet because the action has its source in the integrated consciousness of a *Purusha*, it has to be attributed to him. To say that the *Purusha* is not the actor though things happen round him and are due to him will be seen to be mere quibbling.

It is only in the light of what has been said above that the mystery of Divine action can be understood. The whole universe appears to be the expression of an integrated plan which is working out with extraordinary precision and beauty in every detail. Billions of solar systems grouped in millions of galaxies are serving as theatres for activities of this tremendous Plan, countless hierarchies of beings, both visible and invisible, are serving as its agents in different capacities but the Actor is nowhere to be seen and his actions are equally mysterious and inscrutable. How can such an Actor with infinite ramifications of activities remain unknown and unrecognized and we can trace no activity to its source in Him? Because when action comes out from integrated consciousness it comes out from the consciousness as a whole and not from any particular act of willing as in the case of mind.

This principle is applicable to every level of integrated consciousness. It is applicable not only to a *Purusha* but also to the Logos of a solar system or the Cosmic Logos. Each is supreme in His own sphere and His will is fulfilled in a mysterious manner without any obvious direct indication of its exertion or expression in action. If there is only One Reality and everything that exists in the realm of the manifest and the Unmanifest is an expression of that Reality, it is reasonable to suppose that all that happens in the realm of the manifest is ultimately the expression of the Divine Will inherent in that Reality. The whole of Upanishadic literature is full of unequivocal assertions regarding the supremacy and expression of the Divine Will in manifestation. These statements are made not in reference to an *Īshvara* only but also to *Parabrahman*, the Ultimate Reality, which shows that even the Absolute is not only a Witness but also an Actor. But as no specific state or function can be discerned in that Super-integrated State so no active role can be obviously ascribed to the Absolute. This is so because that Ultimate Reality, as a Super-integrated State, functions *as a whole* and acts *as a whole.*

What has been said above will also throw some light on the significance of the Sanskrit word *Nirīham* which means " without desire ". This word is always used in relation to Divinity at the highest level and not in relation to human beings. For example, the *Dhyāna Mantra* used for meditation on *Parabrahman* begins with the line:

हृदयकमलमध्ये निर्विशेषं निरीहम्

Hridaye kamala madhye nirvishesham nirīham

The use of the word *Nirīham* which means " desireless " in relation to *Brahman* is particularly inappropriate if not absurd, if we take the word in its ordinary meaning. But it acquires a wonderful significance if we take it in the sense of will and action of Divinity which wills but does not appear to will, which acts but does not appear to act, because it wills and acts as a whole. Perhaps this was the original meaning of the word *Nirīham* but later vulgarized by those who did not understand its inner, transcendental significance.

In the above conception of Divine Will and Action is also perhaps hidden the secret of *Nishkāma karma* which is the technique of doing action as an agent of the Divine Life and Consciousness so that no personal *karma* is incurred in the process. In practising *Nishkāma karma* there is not only the difficulty of eliminating personal desires so that we can do what we consider as the Divine Will, but also the difficulty of knowing what is the Divine Will in a particular situation. The Divine Will does not act from the level of the mind and is not governed by ordinary standards of right and wrong and so is extremely difficult to know. The only method available for this purpose is to have the mind attuned to the Divine Consciousness by extreme purification and habitual attitude of self-surrender. The more effectively we can eliminate our personal desires and keep the mind habitually in a pure and harmonized condition the more our consciousness approximates to its integrated state and the more it becomes attuned to the Divine Consciousness. And the more effective this attunement the easier it becomes for us to know the Divine Will intuitively and carry it out effortlessly. In the development and refinement of the faculty of *Buddhi* is hidden the secret both of knowing and doing.

Having discussed the co-existence of the integrated state and the differentiated states derived from it, from the Occult point of view, let us see whether Science can throw some light on this difficult question. We have seen in different contexts that the phenomena of light are in most cases analogous to the phenomena of mind and consciousness and so it will be worthwhile in considering the relation of integrated and differentiated states to examine the nature of light in these two states i.e. as white light on the one hand and coloured lights on the other. For lights of different colours are merely differentiated forms of integrated white light.

Let us first dwell for a while on the integrated and differentiated states of light. The best way of understanding the nature of these two and their mutual relation will be to perform a thought experiment. This is a well-known device used in formulating scientific theories when we have to illustrate a principle by means of an experiment which is theoretically possible but not feasible. Einstein used many such thought experiments in formulating his Theory of Relativity and since then this has become standard practice in discussion of scientific problems.

Suppose we have a globe of white light made by a point light radiating from the centre of a glass globe whose outer surface is silvered. Now we know that white light is composed of seven coloured lights which are seen in the spectrum of white light dispersed by means of a transparent glass prism. So we can easily imagine the globe of white light replaced by seven globes of coloured lights of the spectrum as represented in the following diagram:

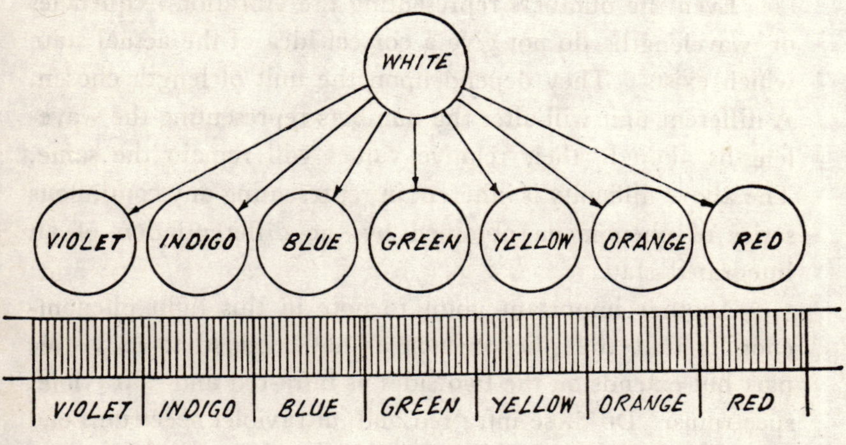

GLOBE OF WHITE LIGHT SPLIT INTO SEVEN GLOBES OF COLOURED LIGHTS

It is easy for any student to imagine the splitting apart of the globe of white light into seven globes of coloured lights, on the one hand, and the coalescing of the separate seven globes of coloured lights into the one globe of white light, on the other. After he does this a number of times he can then imagine the globe of white light as composed of seven globes of coloured lights in one, present and yet not present. All the seven globes are present in the globe of white light potentially but not actually. This is really a case of the co-existence of the integrated and differentiated states.

As a matter of fact this splitting up of the globe of white light into a definite number of globes of coloured lights does not at all give a true picture of the co-existence of the two states. The gamut of vibrations which underlie the visual spectrum is a continuous phenomenon with no breaks or lines of demarcation to separate off one vibration from another. Even the seven different colours of the spectrum merely represent the seven kinds of visual impressions produced on the retina. There is no specific objective change discernible in the series of vibrations when one colour changes into another.

Even the numbers representing the vibration frequencies or wavelengths do not give a correct idea of the actual state which exists. They depend upon the unit of length chosen. A different unit will alter the numbers representing the wavelengths though their relative values will remain the same. The above difficulty is inherent in representing any continuous series of phenomena obtained by the differentiation of an integrated state.

Another important point to note in this light phenomenon is that the phenomenon is not confined to its visible part but extends on the two sides as infra-red and ultraviolet spectrums. Do these infra-red and ultraviolet spectrums end abruptly anywhere or extend *ad infinitum* on the two sides? Our instruments may not be able to detect them beyond a certain range but it stands to reason that they do not end definitely anywhere. If they do not, what happens to them as they continue on the two sides as vibrations of ever-increasing and ever-decreasing wavelengths?

If the student studies the chapter on "Relativity—Total and Partial—IV" he will see how the basis of the spectrum is a continuous phenomenon which can be represented by a circle. It is merely the projection of the integrated state of light as a differentiated state in the lower realm of manifestation. In the integrated state we see the real whole as a circle as it were. In the differentiated state we see it as a polarized phenomenon represented by a straight line extending *ad infinitum* on both sides.

A deeper study of the phenomenon of light thus brings out a very important aspect of the relation between the integrated and differentiated states, i.e., the continuity and indeterminate nature of the differentiated state, which makes it impossible to define and conceive it in a definite manner. It is therefore not only the integrated state which is mysterious and indeterminate, but also the differentiated state which is derived from it.

It is this indefiniteness and indeterminateness at the basis of every natural phenomenon which is reflected in and referred to as the principle of uncertainty which has been discovered by physicists recently. It is this principle which has thrown the whole field of theoretical Science into a state of confusion and uncertainty and paved the way for a due consideration of Occult doctrines in the solution of scientific problems.

This uncertainty is due, as has been explained elsewhere, to the fact that at the basis of the phenomenal world which is governed by exact natural laws is consciousness which is above the laws of Nature. The whole of the manifested universe which is bound by natural laws, and therefore determinate, is embedded, as it were, in consciousness which is free and therefore indeterminate. That is why the uncertainty is found only at the bedrock of any phenomenon and not in the body of the phenomenon itself.

As an integrated state consciousness is free and self-determined. As a differentiated state, mind in its objective aspect is bound by laws of the mental worlds. This mind is a correlate of *Prakriti* which is Nature working according to fixed and exact natural laws. The mind in its subjective aspect is partly an expression of consciousness and therefore partly bound and partly free.

The co-existence of the two states—integrated and differentiated—which cannot be separated from each other shows how the Cosmos is the expression of a Reality which is free in its transcendental aspect and Self-bound by its own laws in its lower aspects. These laws should be considered not as something above and beside the Reality but merely as an aspect of that Reality which finds expression in the objective worlds. This co-existence of the integrated and differentiated states enables the manifested universe to run naturally and automatically and yet to be controlled and guided in its evolutionary process.

It is this fundamental and sacred fact of existence which is symbolized in Hinduism by the *Shiva-Linga* (the solid ellipsoid) resting half immersed on its *Jalahari* (the solid pedestal). The *Shiva-Linga* symbolizes the Eternal and Ever-Unmanifest *Shiva-Shakti Tattva*, the root of Consciousness and Power. The *Jalahari* symbolizes *Mūla Prakriti*, the eternal root of Matter. The manifested universe is an interplay and joint expression of these two eternal and ultimate Principles of existence when they come together.

The freedom and Self-determination of the Reality underlying the universe is indicated by the following well-known Occult maxim in Sanskrit:

चितिः स्वतन्त्रा विश्वसिद्धिहेतुः ।

Chitiḥ svatantrā vishva siddhi hetuḥ

"It is because the Supreme Consciousness underlying the universe is free and Self-determined that the object of the Divine Plan for this universe can be attained with certainty."[1]

[1] *The Secret of Self-Realization* by the author, (1979) p.15, The Theosophical Publishing House, Adyar, Madras.

CHAPTER XVII

THE ROLE OF LIGHT AND SOUND IN MANIFESTATION—I

SINCE the Occult method in the exposition of any profound subject is to start from above and work, step by step, downwards, let us, in discussing the nature of Sound and Light start from their highest aspect in the Unmanifest and come down, step by step, to the lowest aspect on the physical plane. This is the deductive method of investigation and exposition and is possible only in the Occult treatment of subjects because Occultism is sure of its knowledge in its fundamental aspects as embodied in the Occult doctrines. This is so, as has been pointed out on several occasions, on account of the unique method of finding not only secondary truths with regard to the realities in the realm of manifestation, but also the supreme and ultimate Truth of existence which underlies both the Manifest and the Unmanifest. It is not necessary to discuss here this Occult method of finding truth in which there is not only no uncertainty or possibility of error but which is final and all-embracing.

Let us first put down the basic and fundamental facts with regard to the nature and role of Sound and Light very briefly and in the most general terms in the light of the Occult doctrines and scientific theories. This will provide an outline of essential knowledge which can be filled in by further study and research. According to the principle indicated above we shall begin at the highest level in the Unmanifest and come down, step by step, to the lowest level in the manifest, namely, the physical.

(1) The ultimate source of Integrated Sound which is the potential source of all kinds of forces, energies and vibrations in the manifest, is in the unmanifest. Sound is the property of *Ākāsha* and so this highest and ultimate form of Sound must be inherent in *Mahākāsha*. It is this aspect of Reality in its dynamic aspect which is referred to as *Shabda Brahma* and the Logos as " the Word " is only a reflection and expression of this Reality in the realm of the manifest.

(2) On manifestation this Integrated and potential Sound descends through the *Mahābindu* into the realm of manifestation and is then expressed through *Chidākāsha* or mental space created by Divine Ideation. It is from this manifested Sound or *Nāda* that all the energies, forces and motions which lie at the basis of manifestation are derived. This is the realm of the Manifest Logos, the Presiding Deity of a manifested system—Cosmos or solar system—and "Sound" in this realm becomes an instrument of His Will in controlling, regulating and directing all forces and energies in His system.

(3) Though both Sound and Light are expressed as vibrations on the physical plane and serve as instruments of sensuous cognition through the ear and the eye respectively, they play fundamentally different roles in manifestation. This difference is due to the fact that one is the instrument of the positive, static Principle of Consciousness and the other of the negative dynamic Principle of Power, or of the *Śiva* and *Shakti Tattvas*. This " Sound " is symbolized in Hindu symbolism by the well-known symbol of the *Damaru* in the hand of *Mahesha*. The fact that the sound of the *Damaru* is

of the nature of a resonance, produced by two alternate sounds appearing as one, is particularly significant. It hints at the mystery of the role of *Mahesha* in serving as a *via media* between the Unmanifest and the Manifest. It is through His Consciousness that the latent " Sound " in the Unmanifest, referred to generally as *Shabda* becomes potent but integrated " Sound " known as *Nāda* in the realm of manifestation. The latter is also referred to as *Praṇava* and is symbolized by the sound of the conch-shell, a well-known symbol in one of the hands of *Vishṇu* the active Ruler of a manifested system.

(4) Just as Sound is related to Consciousness and is the instrument of Divine Consciousness so Light is related to Power and is the instrument of Divine Power. It provides the material for the objective mechanism of the manifested universe. Modern Science has perhaps provided the most graphic method of expressing this fact by calling matter " bottled-up radiation ". The atom, the smallest unit of matter can be considered, in the light of modern Science as nothing but imprisoned light which is released as free light and other forms of energy on breaking up the atom in atomic fission. This brings to our mind the graphic description in the Bible of the first step in creation in the well-known words " God said, 'Let there be light', and there was light". This obviously hints at the initial creation of the raw material of the manifested universe in the form of light which is then elaborated into discrete particles in the form of atoms of the different planes.

(5) It is in the transformation of light into atoms and their elaboration into all kinds of discrete particles which constitute the basis of the so-called material objective universe that " Sound " comes in. The conversion of light which is initially a radiation travelling freely through space into

atoms in which it becomes locked up in the form of well-known structures is a function of "Sound". It is sound acting from the centre of the atom which provides a field of force in which light in rectilinear motion becomes trapped and transformed into closed circular or elliptical motion. Physicists are puzzled by the incompatible kinds of forces acting within the atom such as the simultaneous presence of forces of attraction and repulsion in the nucleus. They cannot explain these contradictions and in their usual manner have taken them for granted. It is only in the presence of a field of force produced by Sound acting from the centre of the atom that the explanation of these puzzling phenomena can be found.

(6) This kind of role played by Sound is not confined to the atom. It comes into play wherever an organism is working in which different kinds of dynamic forces have to be organized into definite patterns or structures. Thus, the solar system of which the atom is a miniature model is controlled and its movements regulated by means of a field of force acting from the centre of the sun. This field is produced by Sound to which vague and veiled references were made in antiquity in such phrases as "the music of the spheres". The whole universe, with all the organized units within it such as the solar systems, atoms, etc., is vibrating musically. It is uttering rhythmical notes which keep time and thus harmonize the movements and activities of the various units existing within it.

(7) Not only organized units like the sun or the atom which are generally considered inanimate but also living organisms like the human body are kept in an organized condition by means of an inner Sound which creates a kind of field which makes the integration of all the constituent

particles and their harmonious working possible. This Sound proceeds from the "heart" or the *Bindu* in which is centred the life of the whole body and through which descend all the influences and energies from the higher planes into the physical body. This sound is called *Anāhata Shabda* in Hindu mystic literature and can be heard by those who practise *Laya Yoga* and train themselves to hear these inner sounds, and by merging their mind with them, attain higher states of consciousness. According to one of the Upanishads this *Anāhata Shabda* ceases to be heard by the Yogi sometime before his death and this is taken as a clear indication of impending death. This fact is significant and shows that the organized life of the physical body is due ultimately to a Sound and when this sound ceases, death takes place and the constituent particles of the body fall apart.

In considering the role of Sound in the organization of an organism by producing a field of influence in which the organism functions according to its archetype it is necessary to keep in mind a few basic facts if we are to avoid getting involved in all kinds of misconceptions and confusion of ideas regarding these important Occult doctrines.

(a) The first point we have to note in this connection is that the Sound which produces a field of influence for the organism to work in, is not the ordinary sound which we hear through the ear or which we can detect by means of physical apparatus. If this were so, everyone would hear these sounds and this particular role of sound in organizing and controlling an organism would not be a mystery that it is. These subtler sounds which produce these fields of influence are not audible to the ordinary man. Their presence can only be inferred from the effects which they produce in matter. It is only in the practice of Yoga that it is possible to hear these sounds directly and utilize them for manipulating the processes which they control and regulate in the various vehicles of consciousness.

(b) This sound acts on dense matter not directly but through the agency of vital forces which are amenable to the action of mind. This fact can be best understood by taking the case of the human body. It is well known that the dense physical body is permeated by another subtler though still physical vehicle which is referred to as "etheric double" in Theosophical literature and as *Prāṇamaya kosha* in Vedāntic literature. This subtler body is the vehicle of *Prāṇa* or vitality which regulates and coordinates all the chemical, physical and biological processes taking place in the physical body. It is only through *Prāṇa* that mind can act upon matter. This is because of the peculiar constitution of *Prāṇa* which combines in itself the nature of mind and matter and can thus serve as a *via media* between mind and matter.

This also accounts for the fact that sensuous perception requires *Prāṇa* as an intermediary between the dense physical body and the mind. The purely physical vibrations which are received by the sense organs and carried by the nerves to the corresponding centres in the brain can be converted into sensations, which are mental in their nature, only through the agency of *Prāṇa*. This is why in *Pratyāhāra*, the voluntary and complete inhibition of sensation in Yogic practice requires the previous practice of *Prāṇāyāma* which gives complete control over the *Prāṇas* acting in the body.

(c) Etheric matter permeates dense physical matter throughout the solar system and it is through this etheric matter (which is not the same thing as the ether of scientific conception) that the Divine Mind regulates not only the movements of the planets but all other physical activities in the solar system. According to the Occult doctrine the solar system is not a play of blind, insentient forces of so-called Nature but the expression and vehicle of a Divine Being who is referred to as *Īshvara* in Hinduism and Logos in Theosophical literature. The use of the word Logos for the Presiding Deity of a solar system is very significant, for Logos means "the Word" and "the Word" indicates the power of sound through which He controls and regulates all the activities in His solar system.

(d) Science has assumed the presence of many kinds of purely physical forces like gravitation, electro-magnetism etc. in order to account for many physical phenomena like the revolution of the planets, the falling of all bodies towards the earth and other physical phenomena taking place within the atom and between the atoms. Scientists, owing to their materialistic bias, have completely discounted the possibility of mind being involved in any manner in these things. But according to the Occult doctrine the Divine Mind is the chief determining factor underlying these phenomena and what appear as natural forces are really the effects of the action of Divine Mind acting in exactly defined and unchanging ways which we call "Laws of Nature". The God of Occultism is not an old gentleman sitting on a throne somewhere high up in the heavens, ordering about planets and serving as arbiter of human destiny. He is an all-pervading Consciousness which permeates the manifested system which He has created and over which He presides. This Reality expresses itself only partly through natural laws working with mathematical exactitude.

If He were merely to express Himself as Nature and natural laws He would not be very different from the abstract mathematical reality which scientists worship as their god for the time being. He is not only a Reality which pervades the universe in one of His aspects but also a Reality which controls, guides and encloses the universe in His vast, all-embracing love and care. It is this aspect of the Divine Reality with which it is possible for human beings to come into the most intimate relationship of love, and gain help and guidance not only in their worldly life but also in their spiritual unfoldment.

(e) A very unexpected and startling development, which has taken place recently in scientific thought, should open the eyes of those who slavishly accept whatever orthodox scientists propound with regard to the nature of the universe and man. These people are under the illusion that the final word has been said and are prepared to throw overboard even their most cherished religious and philosophical beliefs as superstitions under the inspiration of these scientists. This development has taken place as a by-product of the Theory of Relativity put forward by Einstein and now almost universally accepted by the present scientific world.

Einstein has based his Theory of Relativity on relative motion and accounted for all the phenomena which required the assumption of forces like those of gravitation, electromagnetism on the basis of motion and motion only. He has not only found these forces unnecessary in his conception of the universe but has strongly repudiated the possibility of the existence of such forces which were accepted and taken for granted in the pre-Einstein era of Science. So there is really no such thing as "force" apart from motion although scientists still continue to talk in terms of forces and make their calculations as if these forces do exist. The relativistic conception of the universe without force is not necessarily the Occult conception of the universe, but it has at least removed a major hurdle in the way of accepting the Occult conception. And it has shown how one wrong conception demolishes another wrong conception in the progress of thought so that truth may finally emerge and prevail.

सत्यमेव जयते नानृतम्

Satyameva jayete nānritam

" Verily, it is truth which prevails ultimately, and not falsehood "

We have been dealing so far with the role of Sound in producing, controlling and regulating forms in the phenomenal world. But this is only one aspect of its all-pervading nature. There is another aspect related to consciousness which is far more important to man, though little known except to those who are well-versed in the literature of Yoga. The importance of this role will be seen from the fact that there are two systems of Yoga—*Mantra Yoga* and *Laya Yoga*—which are based on the use of Sound in unfolding higher states of consciousness in man. It is not possible to deal, even briefly, with these two systems of Yoga here but it is necessary to point out the rationale of these systems to show what important role Sound plays in manifestation.

The importance and fundamental nature of this role should be clear from the very fact that the whole structure of the manifested universe is based upon Sound. It has already been pointed out that the Unmanifest Reality has two aspects, one related to the static principle of Consciousness and the other to the dynamic principle of Power. These two aspects are generally referred to as *Brahma Chaitanya* and *Shabda Brahma* or " Reality in its aspect of Consciousness " and " Reality in its aspect of Power in the form of Sound ".

Since these two aspects are inseparable, it is inevitable that when they are reflected, step by step, in the worlds of manifestation at different levels, their mutual relation and dependence should be maintained throughout, though it may not be possible to discern it and trace it in the ramifications of their infinite expressions in the phenomenal worlds. It is this fact which accounts for the intimate association of consciousness and Sound in the worlds of manifestation and for their mutual interaction and capacity to affect each other.

So, when consciousness gets differentiated and descends step by step into different levels of mind, Sound also similarly gets differentiated into less and less subtle states, and at each level of manifestation one can affect the other. It is in this way that mind can utilize " sound " to affect matter and bring about the desired changes in matter if the laws underlying the relation of mind and " sound " and the corresponding techniques are known in a definite manner. It is also because of this fact that " sound " can be utilized in Yoga for harmonizing and reorganizing the vehicles, for affecting the modification of the mind and for unfolding higher states of consciousness. But this aspect of Sound comes within the province of *Mantra Yoga* and *Laya Yoga*.

CHAPTER XVIII

THE ROLE OF LIGHT AND SOUND IN MANIFESTATION—II

WE have already referred to *Nāda* or " Sound " in discussing the mechanism of manifestation and pointed out how all the principles which form the basis of the universe in the manifest state must be present in the Unmanifest as *Shabda* or " Sound ". Since this is a fundamental doctrine of Occultism which holds the secret of the objective universe and can throw considerable light on the theories of modern Science it would be helpful if we consider it in some detail and try to bring it in some kind of relation with scientific knowledge. Although scientists are not prepared to consider seriously the doctrines of Occultism in the conditions prevailing in the scientific world at the present time we should nevertheless try to build and keep ready bridges between Occultism and Science to facilitate exchange of thought when the time is ripe for doing so.

Science has been able to obtain a lot of exact factual knowledge about two physical phenomena—Sound and Light. But it knows very little about their real and essential nature. Occultism, on the other hand, knows the real and essential nature of these two universal phenomena and therefore has not found it necessary to gather so much factual and detailed knowledge concerning their physical nature.

But though there is not so much detailed knowledge concerning the physical aspects of the phenomena of Sound and

Light in Occultism there is enough knowledge concerning their subtler nature to enable Adepts of Occultism to manipulate these subtler forces in a manner of which the men of Science cannot have the slightest conception. Hints regarding the extraordinary manner in which these forces can be manipulated, and the powers which they place in the hands of the advanced Yogi, are contained in some of the aphorisms of the *Yoga-Sūtras* (III-17, III-21, III-22, III-42, I-29). There are still greater powers possessed by the highest Adepts of Occultism about which even a hint cannot be given out for fear of misuse by those who are power-hungry and utterly unscrupulous.

Occultism has not only intimate knowledge regarding the subtler aspects of Light and Sound and can therefore manipulate these forces effectively from within, but what is more important, it has the transcendental knowledge concerning the fundamental role which Light and Sound play in the mechanism and process of manifestation. It is this knowledge and the capacity to utilize it for practical purposes which confer Omnipotence on the highest Adepts of Occultism and enable Them to control and guide effectively the manifested systems over which They are required to preside. Real and complete knowledge and effective power go together. This principle is seen working in its very limited physical aspect even in the field of scientific work.

Before considering the subtler and essential nature of light and sound which are merely the joint expressions on the physical plane of the highest Principles of Power and Consciousness in the Unmanifest, it is necessary to clarify once again our ideas regarding the essential nature of the Divine Power which not only provides the mechanism of the objective universe but also enables consciousness to express itself through this mechanism. There is a great deal of uncertainty and confusion with regard to the real nature of the Divine Power which is generally referred to as *Shakti* in Hindu schools of philosophy. Some schools of philosophy go to the length of considering it as identical with the Ultimate Reality, while others regard it merely as the active agent behind the insentient forces and energies which produce natural phenomena. As a matter of fact both these views are different aspects—the highest and lowest—of the same Reality, and it would help us a great deal in understanding the Occult doctrines clearly if we can mentally grasp how these two extreme views can be reconciled even from the logical point of view.

In order to understand how the same Reality can exist in these extremely different and apparently irreconcilable forms it is necessary to go back in thought to the highest and the most fundamental doctrine of Occultism according to which there is One and only One Reality in existence. If there is only One Reality—all-inclusive and all-pervading—and nothing except that Reality can exist, manifestation of any kind derived from that Reality must, by its very nature, be a change within that Reality Itself, as there can be nothing outside It. It should also be obvious that when a change of any kind takes place in such a Reality it cannot be essentially a change from one kind of thing to another kind of thing as we consider it in the realm of manifestation. It must be a change from one aspect or state of that Reality to another aspect or state of the same Reality.

Now, if a change of this kind is produced it can be produced only by motion of some kind in that Reality. This is a subtle idea and in order to grasp it we can take the help of a concrete illustration. Let us try to imagine some kind of change in pure water which keeps it exactly in the same condition as a substance as well as in quantity and quality. However hard you try, you cannot produce or even visualize any change which is anything but motion of some kind within the substance of the medium. Motions of different kinds within the medium of such a substance is the only kind of change which is possible under the conditions referred to above. So when manifestation takes place in the Ultimate Reality, which is One, that manifestation can basically be only of the nature of different kinds of motions taking place in that Reality—nothing but motion.

A closer analysis of the conditions which will be produced by different kinds of motions in pure water will show that the things which appear as a result of these different kinds of motions will be of an illusory nature. By "illusory" we mean that they will not be what they appear to be. Let us consider more critically the conditions present in pure limpid water at rest in which different kinds of motions are introduced in order to produce what appear as different kinds of forms which are phenomenal and cognizable. Let us imagine that the water is churned by a turbine in different ways to produce an infinite variety of patterns. This churning will produce a constantly changing phenomenal world of forms in a medium which in itself is free from change.

Let us now imagine another kind of change in internal molecular motion by heating the water on one hand and cooling it on the other. Heating will convert it into water vapour which has the properties of a gas and cooling it sufficiently will convert it into ice which has the properties of a solid. The important point to note in all these changes of state produced by motions of different kinds is that the water will remain water as a basic substance. The changes are merely changes of appearances with absolutely no change in the substance or the nature of the underlying reality. And such a change, if we think carefully, is purely illusory in the strict sense of the term "illusory".

The above concrete experiment provides a very apt illustration of the Oneness of the underlying Reality amidst the infinite variety of forms and states found in the phenomenal worlds. These definite forms and states are produced by Self-initiated movements of different kinds within the Reality called *Spandana* in Sanskrit, the Reality itself remaining unchanged in its essential nature of pure Consciousness. The fact that nothing else except motions of different kinds are involved in the production of these different forms and their movements, also makes it quite obvious that the phenomenal world is illusory in the sense that it is not what it appears to be—as something different and independent of the Reality from which and in which it is produced by motion.

If the student has understood the underlying deep significance of what has been said above, he will see how futile the controversy between the two schools of Vedāntic philosophy, referred to as *Vivarta Vāda* and *Māyā Vāda*, is. The first regards the phenomenal manifest universe as a product of *vivarta* or modification in *Brahman* or the unmanifest Reality. The second school of *Māyā-Vāda* regards the phenomenal universe as a pure Illusion or *Māyā*, the underlying Reality remaining unchanged as the rope remains unchanged when it is taken for a snake in the dark. As a matter of fact both points of view are true when the phenomenal universe is looked at from different points of view as in the case of water in which an infinite number of forms can be produced by merely churning it. There is a kind of change produced in the underlying Reality but this change is merely a motion in the substance of Reality and therefore no real change.

The phenomenal world produced by these motions of different kinds is merely an appearance and therefore "illusory" because the underlying Reality remains unchanged. Even the illusory nature of the phenomenal world does not represent the whole truth because the forms are produced in the substance of Reality and though constantly changing are essentially of the nature of that Reality. It is this fact which has made certain Tāntrik philosophers deny the illusory nature of the phenomenal universe in the following significant statement:

न मिथ्यात्वाय कल्पन्ते जगत: शिवरूपत:

na mithyātvāya kalpante jagataḥ Shiva-rūpataḥ

"The universe cannot be considered as an Illusion because it is a form assumed by *Shiva* the Supreme Reality."

This paradox created by the intellect is resolved only in the vision of Reality in Self-realization when the phenomenal nature of the manifested universe is realized and yet the whole phenomenon is seen as a play of Divine Consciousness as expressed in the Occult maxim:

ब्रह्मैवेदं विश्वम्

Brahmaivedaṃ Vishvaṃ

"Verily, this universe is nothing but Brahman."

The fundamental doctrine of Occultism discussed above provides very briefly the background and broad basis of the whole Occult philosophy. All other doctrines must conform to it and may be considered merely as elaborations and different aspects of the central doctrine in different spheres and levels of existence. The more deeply we enter into the inner significance of these doctrines the more clearly we see how they are intimately related to one another and rooted in the fundamental doctrine of One Ultimate Reality. Like the roots of *Ashvattha* or the Tree of Life hanging downwards from the Unmanifest this doctrine supports and is the central Truth underlying the whole Occult Doctrine. Indeed, it would be an interesting mental exercise for a student of Occultism to make an attempt to deduce, step by step, the whole Occult Doctrine in its broad and essential outline from the central doctrine of the One Reality. He may not succeed fully in doing so but he will at least gain a partial realization of the fact that these different doctrines are intimately related to one another and like the abstract truths of mathematics are inherent in the very nature of existence.

Coming back to the problem under discussion, and accepting the broad principle that Motion in the abstract is the basis of manifestation and of the phenomenal worlds produced in the One Reality as an appearance, let us now go into a little more detail and consider the nature of this Motion, the various forms it can assume on differentiation and the different ways in which these can combine in order to produce the infinite variety of phenomena in the manifested worlds. Our ideas about Motion and Energy are derived from our experiences of the physical world and since the physical universe is only the outermost shell of the total universe which exists on the different planes, naturally, these ideas must be very limited and partial. So when we try to understand how Motion can account for and serve as the basis of the manifested world we must be conscious of our great limitations and not hope to grasp the realities underlying this Occult doctrine except in a hazy outline.

It is also necessary to say a few words with regard to the use of the word "Motion" to indicate in a general manner the dynamic aspect of the Reality underlying the phenomenal world. There are different aspects of this dynamism for which different words are used in scientific work such as "force", "energy", "work", "power", etc. These different aspects are related to one another in a definite manner. For example, "power" is defined as work divided by time, "work" is defined as force multiplied by distance. Forces are measured by comparing them with the force with which bodies are attracted by the earth, the phenomenon which is referred to as gravitation.

Since all these quantities are measured ultimately in terms of motion and there is a great deal of uncertainty with regard to their real nature in scientific circles, it is better to confine ourselves in the present context to the word "motion". This is a cognizable phenomenon which everybody can understand. What produces motion is not only debatable but difficult to understand.

That this is not only the simplest but also the most appropriate method of indicating the dynamic aspect of the basis of the manifested universe will be seen when we take into account the Occult doctrine concerning the basis of the phenomenal worlds. According to this doctrine these worlds are essentially a mental phenomenon and these phenomena are produced by motion of one kind or another in mental stuff. We need not go into this question here because it has already been discussed in some detail in appropriate contexts, but the student would do well to try and grasp clearly the general principle underlying this Occult doctrine if he wants to understand the role of "Sound" and "Light" in producing the phenomenal world.

An added interest has been imparted to this Occult doctrine on account of the Theory of Relativity. Einstein threw overboard many of the well-established theories of modern Science which were till then universally accepted and accounted for some of the most fundamental phenomena of the physical world, and he accounted for these phenomena on the basis of motion and motion alone. It is true that the way in which motion has been made to account for gravitation and other natural phenomena in the Theory of Relativity does not agree with the Occult conception of the role of motion in producing the phenomenal world, but still, it is definitely a step in the direction of the Occult doctrine. If both the Occult doctrine and the scientific theory, assign a fundamental role to motion in the production of natural phenomena and are agreed on the general principle, the manner in which motion accounts for the phenomena becomes a secondary question and can be expected to be sorted out and answered later as further developments in different fields of Science take place and show the fallacies and inadequacies of the scientific theories.

It is not possible to discuss in detail the nature of Sound and Light and their roles in producing the phenomenal worlds. This is too vast and too complicated a subject, and too little has been divulged to the general public about the subtler aspects of these two natural forces, to enable us to have a reasonably clear and complete picture of the two phenomena. We shall therefore merely refer briefly and in general to a few fundamental facts concerning the nature and roles of Sound and Light in manifestation.

We should, however, keep in mind what has been said about *Nāda* and its differentiation into an infinite variety of forces, energies, motions, etc., which perform their respective roles in the mechanism of manifestation. They not only perform different roles but each force, etc. becomes transformed into more and more dense and less and less effective forms as it descends, step by step, into the lower levels of manifestation, so that there may appear hardly anything in common between the highest and lowest forms except the essential nature of the principle.

But though in these extremes, highest and lowest, forms and aspects differ so greatly in outer appearance, there is an indissoluble thread of common essential nature running between the different aspects and forms of the principle and they together constitute an integrated, harmonious and complete whole. The *Ashvattha* or the Tree of Life is thus not only a symbol of manifestation as a whole, showing the differentiation of the Ultimate Reality into different Principles or *Tattvas*, but also a symbol of each individual principle, showing its differentiation into different forms and aspects as it descends lower and lower into the denser levels of manifestation. This must be so if manifestation at different levels is a repeated projection, pervasion and expression of the One and the same Reality into more dense and less real states of Itself.

This is how Sound and Light which are Cosmic Principles in their highest aspects descend lower and lower into manifestation in less and less subtle forms and ultimately assume the form of physical sound and light, with which we are familiar, and which physicists have investigated with such thoroughness. But as the outermost shell and crudest form of the *principles* have been investigated, however exact and thorough their investigation, they cannot get at the real nature or even at the subtler forms of the principles, though they may be able to know a great deal about their physical forms and be also able to control and manipulate them for their extremely limited physical aims. No wonder the essential and real nature of Sound and Light has remained a baffling mystery to the modern scientist in spite of all the scientific developments which have taken place in physics.

CHAPTER XIX

THE ROLE OF LIGHT AND SOUND IN MANIFESTATION—III

We have already referred in a previous chapter to the essential nature of Light and its fundamental relation to Sound in the mechanism of manifestation. The essential nature of Light is radiation and as matter is nothing but "bottled-up radiation" Light may be considered as the raw material of the universe. Of course, the word radiation is used in Science to indicate all kinds of electromagnetic vibrations, visible or invisible, and the use of the word light is restricted to those vibrations which can affect the eye. But as the layman is not expected to be familiar with the distinctions between these different kinds of vibrations it is permissible for him to use the more familiar word light as a representative of all kinds of electro-magnetic vibrations.

The relation of Light to Sound is of a very fundamental nature, though in understanding this relation we cannot get much help from the investigations made in Science. The reason for this lies, as has been pointed out before, in the fact that Sound is related to and is an instrument of Consciousness while Light is related to and is an instrument of Power. The whole objective universe which, as has been shown elsewhere, is nothing but condensed light, is a play of Power or *Śakti* and is a visible, apparently tangible phenomenon which can be investigated by instruments which are also in the realm of the phenomenal world. It has therefore been easier to investigate the phenomena of light by scientific methods. Consciousness, though the source of Power, is the subjective aspect of Reality and is always hidden, and present in the background of the phenomenal world. It cannot be perceived or investigated by the objective methods of Science, belonging to the phenomenal world. It can be perceived and investigated only by the subjective methods related to the Real world of Consciousness. Yoga, which is based upon and utilizes these subjective methods of investigation, is therefore the only correct and available method of investigating and knowing the realities of the Real world which are in the realm of Consciousness.

Since Sound is related to and is the instrument of Consciousness, the subjective aspect of Reality, its functions in the realm of manifestation are not primarily of an objective nature and not amenable to investigation by the objective methods of Science. It is only the repercussions of Sound in matter which appear as physical vibrations which can be cognized by the ear and detected by means of physical instruments. It is only these which have been thoroughly investigated by Science.

The subtler aspects of Sound which are generally referred to as *Vaikhari*, *Madhyamā*, *Pashyanti* and *Parā* have been investigated only by Occult methods. They provide means of communication between different states of consciousness and form the basis of esoteric techniques of *Mantra Yoga* and *Laya Yoga*.

In trying to understand the essential nature of light and its role in manifestation let us begin with a simple scientific experiment which every school boy has to perform when he studies the nature of electricity and magnetism.

If a bar magnet is thrust into a coil of metallic wire an electric current is produced in the wire as shown below:

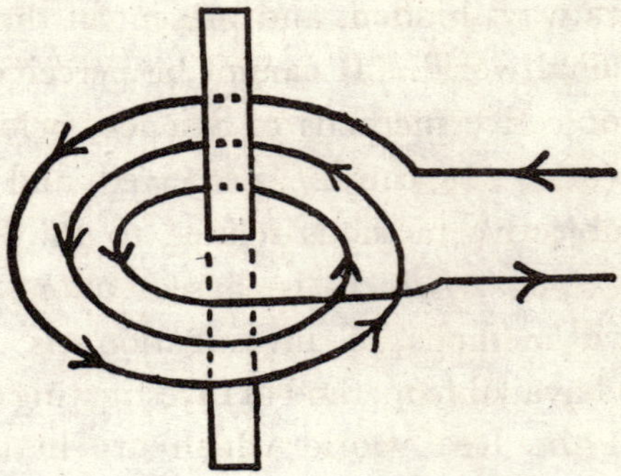

The above experiment shows that a polar dual principle containing positive and negative aspects in a potential form is capable of producing energy in a kinetic form. Two things needed for this purpose are (1) motion or existence of force (2) a mechanism like a conducting wire which can serve as a vehicle of the kinetic energy generated.

The natural process embodied in the above experiment will be seen to reflect in a remarkable manner the process by which manifestation is brought about from the Unmanifest. The *Śiva-Śakti Tattva* is the ultimate basis of the manifested universe and remains in contact with *Prakriti* in its *Sāmyavasthā* or potential state in the Unmanifest. When the time for manifestation comes after a period of *Pralaya* the Divine Will moves, the harmonized condition of *Prakriti* comes to an end and the *Guṇas* begin to work. In the case of the scientific experiment it is the exertion of force which produces the movement of the bar magnet and generates the electric current in the wire which can be converted into different forms of energy. In the case of manifestation it is the exertion of the Divine Will or

Force which energizes the *Prakriti* or Nature and produces *Nāda* the potential force which takes infinite forms of motions, vibrations and movements in the realm of manifestation. The mechanism of *Prakriti* in a potential state which is called *Mūla Prakriti* is eternal and is always there to receive the Divine impulse and become energized into action.

It is interesting to note that both the positive and negative forms of magnetism are present in the bar magnet in an invisible mysterious form. All we can see from outside is the outer vehicle of the dual force, namely, the bar of steel. Again, when the magnet moves and produces the electric current in the wire, electricity the primary form of energy is present in an invisible mysterious form and only the outer vehicle of the electric current, namely, the iron wire is visible. So, outwardly, it is only a movement of matter in one mechanism in relation to matter in another mechanism and yet all these subtle invisible forces are working within these mechanisms and can be cognized by the different kinds of effects which they can produce in physical apparatus designed to detect them. So, let not scientists say that the subtler forces of Nature do not exist simply because they are invisible and their crude physical apparatus cannot detect them.

Since mere motion of an unmagnetized bar of steel cannot produce electricity in the wire it follows that the source of electricity is in the magnetism of the steel bar in an invisible mysterious form and it is this which produces the electric current. Similarly, the polar *Śiva-Śakti Tattva* is the ultimate cause of manifestation and the ultimate source of *Nāda* and all the energies and vibrations which are derived from it by a process of differentiation. The positive element of Consciousness in this dual Principle is the source of the omnipotent Divine Will which remains hidden in the background and the negative element of Power is the source of the infinite Divine Energy which appears in manifestation and produces the phenomenal world.

In the case of a bar magnet which represents a polar phenomenon the force of positive and negative magnetism is maximum at the extremities which are therefore called poles of the magnet. At the centre the two opposite forces are balanced and the magnetism is therefore zero. So that if we move from the positive pole to the other pole the positive magnetism decreases gradually, becomes zero, the negative magnetism then begins to increase and becomes maximum at the negative pole as shown below:

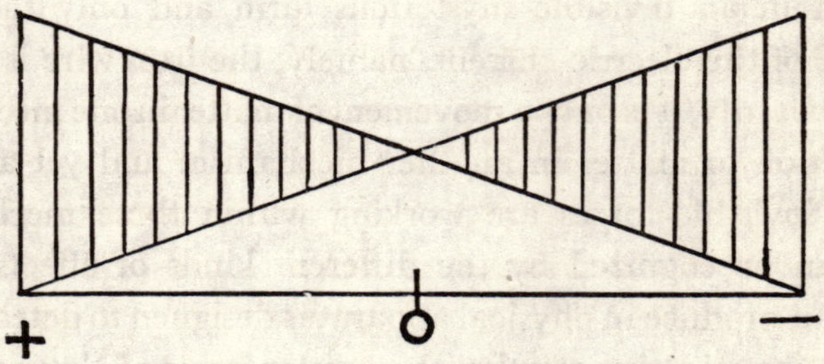

These conditions in a bar magnet reflect and illustrate in a very remarkable manner the relation of the two polar Realities in the *Śiva-Śakti Tattva*. *Siva* and *Śakti* or Consciousness and Power, being polar Principles, are inseparable and interdependent but one can be more predominant than the other in the realm of manifestation. It is only in the Unmanifest that they exist in a perfectly balanced and harmonized State as the Absolute. In the realm of manifestation one is more predominant than the other and it is this fact which determines whether the particular Reality appears as an expression of consciousness or its differentiated form, mind, on the one hand, or as an expression of energy or its condensed form, matter, on the other. In both cases one is in the forefront and the other in the background. In both cases the pure integrated form is one while the differentiated forms are many. Pure integrated Consciousness of *Siva* is one, its differentiated forms in mental expressions are infinite in number. Pure integrated Power of *Sakti* and even pure integrated Energy as *Nāda* is one while its differentiated forms present in different kinds of vibrations, atoms and molecules are infinite in number.

As every natural phenomenon is a joint expression of two opposite polar principles which reflect and express at different levels the primary polar principles of Consciousness and Power let us see whether it is not possible for us to trace or discern the presence of both the principles in such phenomena. Let us take an electric current. Here it is the dynamic element of electromagnetism namely electricity which is predominant and can be detected easily. Where is the static element of magnetism in the electric current? It cannot be seen but it must be present in the electric current. How do we know? When we pass a current in an electric fan, a magnetic field is produced in the field coils and it is this which makes the fan revolve. This is the principle underlying every electric motor. Where does the magnetism in such a case come from? From the electric current, of course. If we pass a current of water or any other fluid in a coiled tube it would not produce a magnet. There is no other means of producing a temporary magnet in this way and there can therefore be no other source of this temporary magnetism except the temporary electric current which produces it. So, it is obvious that an electric current carries within it not only the electric force but also magnetism which produces a magnetic field round the wire.

We have seen already how an electric current is produced in a wire by thrusting a magnet in a coil of wire. Where is the source of the electric current in this case? It is obvious that this source must be in the magnetism of the magnet because the movement of an un-magnetized bar of steel in the coil cannot produce an electric current. So the above experiments and the conclusions based upon them show conclusively that electricity and magnetism contain each other in a mysterious manner though we cannot easily detect any electricity in a magnet or magnetism in an electric current.

If we shift our attention from the electromagnetic phenomenon to the polar principle of Consciousness-Power we shall find not only a similar relation existing between the two but also a similar action of the one on the other. We have already seen in a previous chapter that according to Occultism every force has Consciousness in the background. All forces of Nature which bring about all kinds of movements and changes in the phenomenal world are *conscious* forces. Science naturally cannot see this element of mind or consciousness because it utilizes only physical means. Occultism investigates these natural phenomena not superficially like Science but in a total manner by means of mental faculties, and can therefore perceive the element of mind or consciousness behind the apparently insentient forces or powers working in Nature.

If all natural forces and powers are guided and motivated by Consciousness in the background, is the reverse of this also true? Yes. All expressions of mind and consciousness have power or the capacity to exert force hidden in the background even when they are in a state of repose. Every mental faculty, every expression of the mind in some kind of specific knowledge has a corresponding power associated with it as we can see in our ordinary daily life and as the technological developments in Science have demonstrated in such a spectacular manner. Extraordinary knowledge with respect to the finer forces of Nature working on the subtler planes which is gained in the practice of Yoga confers extraordinary powers on the Yogi as the third chapter of the *Yoga-Sūtras* shows very clearly. But while mental faculties confer only mental or psychic powers, pure consciousness confers spiritual powers like *Ashta Mahā Siddhi*. And ultimately, Omniscience confers Omnipotence on the Yogi. So, not only does every power contain consciousness in the background but every expression of consciousness has its corresponding power associated with it either in a potential or active form.

The remarkable similarity between the highest polar principles of Consciousness and Power and phenomena of a polar nature on the lowest planes is not, according to Occultism, accidental or a mere coincidence, the ever-ready explanation of the sceptic. It is not even a mere ' reflection ' as indicated by the Occult maxim "As above, so below." It is the direct result of the two most important facts of existence embodied in two fundamental doctrines of Occultism. These are (1) There is only One Ultimate Reality in existence and this manifested universe is the result of the repeated projection, reflection and expression of that Reality in progressively lower states of Itself. (2) The manifest universe is the joint play of two polar ultimate Principles in existence which are the root of all expressions of consciousness and power and are referred to as *Śiva* and *Sakti* in Hindu philosophy. These two Occult doctrines are embodied in the two following Sanskrit maxims.

Brahmaivedaṃ Viśvaṃ

"Verily, this universe is nothing but *Brahman* the Ultimate Reality"

Śiva-Śaktyātmakam Viśvaṃ

"*Śiva* and *Śakti* are the innermost Soul and Essence of the universe."

It is necessary for the student of Occultism and Science to understand, as far as this is possible in the realm of the intellect, the nature and relation of these two highest Principles of Consciousness and Power and how they manifest in the lower worlds and by their mysterious joint action produce the phenomenal worlds against the background of pure Consciousness. It is this pure Consciousness of *Śiva* which like a perfect polished mirror reflects every thought in Divine or individual Ideation and produces the illusory impression of real worlds. It is this ultimate Power or *Śakti* which differentiates and through the *Three Guṇas* provides the objective basis of the phenomenal worlds. When a world of form, colour and motion is produced in a polished mirror in this manner there is nothing in the visual phenomenon except the imageless polished mirror on the one hand and vibrations of light which produce the visual images on the other. Similarly, when these phenomenal but apparently real worlds are produced in consciousness there is nothing in existence except pure Consciousness which serves as the background of the mental images and the light of consciousness by which the images are produced and perceived. The mental images themselves which are perceived will be seen on deeper thought to have no real existence but to consist merely of forms which are temporary impressions produced in Consciousness and therefore illusory. Even if we consider them to have a quasi-reality of a sort their stuff is mental as mind is merely a differentiated form of Consciousness.

The above simile brings out the nature of light in its highest aspect as light of Consciousness through which the objective world of mind is perceived. This light of Consciousness, though essentially one, also appears to be of many aspects owing to its different functions and spheres of action. This fact is expressed very beautifully in the following Sanskrit *śloka* in which an attempt is made to give some idea of the different forms or functions of that Supreme Consciousness which is referred to as *Śiva* and which is the only Reality in existence.

अन्तर्ज्योतिर्बहिर्ज्योतिः प्रत्यग्ज्योतिः परात्परः ।
ज्योतिर्ज्योतिः स्वयंज्योतिरात्मज्योतिः शिवोऽस्महम् ।।

antarjyotir bahirjyotiḥ pratyagjyotiḥ parātparaḥ
jyotirjyotiḥ svayaṃjyotirātmajyotiḥ Śivo'smahaṃ.

It is impossible to convey the profound meaning and significance of this *Śloka* in a few words but as it gives some idea of light in its highest aspect let us at least consider it briefly. Let us first give the literal translation of the Sanskrit words and phrases used in the *śloka*:

अन्तर्ज्योतिः	=	Internal light
बहिर्ज्योतिः	=	External light
प्रत्यग्ज्योतिः	=	Inward-turned light of spiritual consciousness
परात्परः	=	Transcendental (which is)
ज्योतिर्ज्योतिः	=	The Light of lights
स्वयंज्योतिः	=	Self-illuminated Light
आत्मज्योतिः	=	The Light of *Ātmā* or the Spirit
शिवोऽस्महम्	=	I am that *Śiva*.

"I am essentially that *Shiva* whose essential nature is the Light of Pure Consciousness which illuminates Itself, is the Real illuminator behind all kinds of lights, which is that transcendental Light of Reality which also expresses Itself in external and internal lights."

Let us now try to understand the inner significance of the different aspects of light referred to in this *śloka*.

The first two phrases 'internal light' and 'external light' refer to the kind of light which is the instrument of perception through a sense-organ. On the physical plane which is the lowest, visual perception of objects takes place with the help of light with which we are familiar and which Science has investigated so thoroughly by means of physical apparatus. The organ of perception called an *Indriya* in Sanskrit is the eye in this case and the instruments of perception are the electro-magnetic vibrations which strike the retina of the eye and produce nervous impulses. These are carried to the corresponding centre in the brain and are converted by a mysterious process into sensation. We need not go into this question here as the mechanism of sense perception is well known and has been discussed in other contexts. The phrase *Bahirjyotiḥ* obviously refers to this physical light which illuminates objects in the external physical world.

As is well known to students of Occultism there are subtler worlds of finer matter and vibrations hidden within the physical world. On the next two subtler planes, called astral and lower mental in Theosophical literature, perception still takes place with the help of sense-organs though the mechanism of sense-perception is not exactly the same as on the physical plane. Not only is a sense-organ necessary but an external agent like light to stimulate this sense-organ is also necessary. The nature of this instrument has not been investigated scientifically, but considering the relation of these worlds with the physical it is presumably light of smaller wavelengths which is invisible to us but can illuminate objects on these subtler planes. Since these subtler bodies exist within the physical body and are centred round the point of consciousness referred to as the *Bindu*, it is possible to stimulate the sense-organs of these bodies and also make the physical body sensitive to the vibrations of these subtler planes. Those who succeed in doing so can perceive the phenomena of these subtler planes also as we all perceive the phenomena of the physical plane. Their mental horizon becomes enlarged and embraces the next two planes also. The phrase *antarjyotiḥ* refers to light as it exists on these inner subtler planes. It is essentially of the same nature as physical light but as it illuminates objects of the inner planes it is called ‘ Internal light ’.

In order to understand the significance of the next phrase *pratyagjyotih* we should recall that perception of the spiritual planes of the Individuality, which functions on the Higher Mental, Buddhic, and Lower Atmic planes, is of an entirely different nature from that on the lowest three planes referred to above. The reason for this, as has been shown in other contexts, lies in the fact that the vehicles functioning on these planes are ‘ atomic ’ in nature, and not like the ‘ bodies ’ used by the personality on the lowest three planes. It is not necessary to enter into this question here, firstly, because it is not possible to visualize the conditions of existence on these spiritual planes and secondly, because the matter has been discussed in other contexts. But we should try to grasp the general principle that consciousness functions in a fundamentally different manner on the spiritual planes of the Individuality and this aspect of consciousness is referred to as *pratyak chetanā* in Yogic terminology. If the student reads the commentary on aphorism I-29 of the *Yoga-Sūtras* he will be able to gain at least a vague and general idea about this *pratyak chetanā*.

Pratyagjyotiḥ is related to *pratyak chetanā* and may be translated as the light of spiritual consciousness. On the spiritual planes there is a mysterious relation between light and consciousness and the more we penetrate into the deeper realms of consciousness the more difficult it becomes to distinguish it from light. So, the former may be considered as the same reality in its aspect of light and the latter in its aspect of consciousness.

We cannot know or visualize the nature of this spiritual light on the lower planes but there is one aspect of its nature, which we may note here. It is capable of expression and unfoldment to an infinite extent. This capacity for unlimited expansion is found even in its expression on the physical plane in the form of ordinary light. Physical light continues to expand and travel through empty space of the universe for millions of years and this process never comes to an end theoretically. The expansion of the spiritual light of consciousness is of a different nature, but still, we can see in the above physical phenomenon a remarkable resemblance to the nature of the spiritual light of consciousness.

The capacity for infinite expansion which is inherent in this spiritual light of consciousness is hinted at in aphorism II-28 of the *Yoga-Sutras*.

" From the practice of the component exercises of Yoga, on the destruction of impurity, arises spiritual illumination *which develops into awareness of Reality*."

But a more detailed, graphic and beautiful description of it is given in *Light on the Path*, especially in the Commentary on aphorism 20 of Part I.

"Seek It (the Reality) by making the profound obeisance of the soul to the dim star that burns within. Steadily, as you watch and worship, its light will grow stronger. Then you may know you have found the beginning of the way. And when you have found the end its light will suddenly become the infinite light."

The above paragraph will show the inner significance of the phrase *parātparaḥ* and why this light is called 'transcendental'. It is true that this spiritual light in its earlier stages of unfoldment is within the realms of manifestation and therefore more or less obscured by the veils of the mind. But it has the inherent capacity to unfold *ad infinitum* and merge into the Light of Reality or the Reality Itself, for, in that State the triplicity of Perceiver, Perceived and Perception has become the unity of undifferentiated Consciousness or Reality.

In the second line of the *śloka* the higher aspects of Light as Pure Consciousness are referred to. *Jyotirjyotiḥ* means the Light of lights. The phrase obviously refers to the Light of Consciousness. Whatever the mode or mechanism of perception, consciousness is always present in the background, and as has been explained elsewhere, provides the illuminating power of perception at every level. In *Pratyāhāra*, when consciousness withdraws from the sense-organs, even though the eyes are open nothing is seen. In *Śāmbhavi Mudrā* referred to in the following *śloka* the same conditions are present.

अन्तर्लक्ष्यं बहिर्दृष्टिर्निमेषोन्मेषवर्जिता ।
एषा सा शाम्भवी मुद्रा वेदशास्त्रेषु गोपिता ।।

antarlakshyam bahirdrishtirnimeshonmesha varjitā
eshā sā Śāmbhavī mudrā vedasāstreshu gopitā

"When the consciousness is turned inwards towards an inner object (of contemplation) but the eyes remain open without blinking and yet seeing nothing, the state is called *Sāmbhavī Mudrā* and is considered a mystery by the scriptures."

Śvayamjyotiḥ which means 'Self-illuminated Light' is that power of perception in Consciousness by which it is able to illuminate itself and perceive itself in Self-realization. This mode of perception can best be illustrated by means of a central light surrounded by a number of semi-transparent glass globes. The light can penetrate through all the globes and illuminate them all. If all the globes are removed one by one and only the central light is left, it illuminates itself and reveals itself in its true nature and full brilliance.

This process of unveiling and ultimately illuminating or revealing itself of consciousness is referred to in aphorism IV-22 of the *Yoga-Sūtras*. In ordinary perception through the mind, the Perceiver, Perceived, and Perception constitute a triple state. In Self-perception of Consciousness by itself the triplicity becomes merged into one integrated state. It is this fact which makes the awareness of Reality, which is an integrated state, possible. Only like can perceive like. It will be easy to understand the nature of Self-perception if we remember that in the triplicity of perception the object of perception and the faculty of perception which is called *Buddhi* both come out of the Perceiver, or to put the same idea in philosophical language, the *Chit* and *Ānanda* aspects are inherent in the *Sat* aspect and are derived from the *Sat* aspect in triple perception.

This Integrated State of Reality in which the faculty of perception is merged in the Perceiver is referred to as *Ātmajyotiḥ*. The Spirit is centred in Itself and sees Itself by Itself if we may put it in this enigmatic form. This is *Kaivalya*, the goal of Yogic discipline. It is called *Kaivalya* or ' Isolation ' because in that state the *Purusha* is completely isolated from *Prakriti*, Power has become completely merged in Consciousness and Pure Consciousness or the *Śiva* State alone remains. It is this State in which every human being exists in his innermost Self and which he is destined to realize in an ever increasing measure in the long course of human and superhuman evolution. And it is this union or the mergence of the individual consciousness in the Supreme Consciousness of *Śiva* which is referred to in the last phrase—*Śivo 'smaham*— in the second line of the *śloka*.

CHAPTER XX

THE ROLE OF LIGHT AND SOUND IN MANIFESTATION—IV

WHAT has been said in the previous chapter should show that Light like Sound is a fundamental Principle which pervades the whole field of manifestation. It plays a triple role in manifestation. It provides the raw material of the objective world as we have seen in the case of the physical world. For, matter according to modern Science is nothing but condensed light or radiation. It is also the instrument of perception for we see objects with the help of light. As manifestation is the result of a repeated projection of the Ultimate Reality into less and less subtle states which appear as matter we may expect that the perceptive role of light becomes more and more predominant and important as we penetrate into deeper levels of consciousness. This is actually so, as we have seen in the last chapter. As consciousness sinks into deeper levels of Reality in the practice of Yoga, Light, the instrument of perception, not only becomes more and more subtle but less and less distinguishable from Light the Perceiver. And in the final stage of Self-realization Light the instrument of perception merges with the Perceiver in an Integrated State in which the two are indistinguishable.

Let us leave for the present the subtler aspects of Light which we have been considering and dwell for a while on its lowest aspect in the form of electromagnetic phenomena which have been studied thoroughly by Science. It is necessary to begin consideration of this aspect of light by recalling that there is a remarkable similarity between the behaviour of Consciousness and that of physical light. The student of Occultism who has studied the Occult doctrines in the light of discoveries made in different fields of modern Science cannot fail to be struck by the analogous behaviour of the two. These similarities of behaviour have been pointed out again and again in *Man, God and the Universe* and *The Science of Yoga* and if we study these facts carefully it is impossible to avoid the conclusion that there is a very intimate and fundamental though mysterious relation between Consciousness and even physical light.

Scientists have not noticed and know nothing about this relation because they have studied only physical light and do not recognize, at least officially, the existence of consciousness as an independent principle in existence. So the question of noting this analogous behaviour of light and consciousness and taking into account its significance in the conception of the nature of the universe does not arise in their case. The student of Occultism has a great advantage over the scientist in having a fuller and truer conception not only about the nature of light but also of other purely physical realities. He has at his disposal not only the vast and detailed knowledge of Science but also the more fundamental and reliable knowledge of Occultism incorporated in the basic doctrines of Occultism. This knowledge helps him not only to view the scientific knowledge in its true significance and correct perspective but also to integrate it with the body of Occult knowledge for a clearer understanding of the nature of God, man and the universe.

The intimate relation of consciousness and physical light suspected from the remarkably analogous behaviour of the two has been stated in very explicit terms in the Sanskrit *śloka* quoted in the previous chapter. For, according to this *śloka*, the different forms or aspects of Light mentioned in it are merely expressions of that Supreme Consciousness which lies at the basis of the universe and is also the innermost Reality hidden within the heart of every human being. We may not be able to correlate these expressions with this Consciousness or comprehend their nature but the fact is there and we can utilize it in formulating a richer conception of the nature of man and the universe.

It has already been pointed out in the previous chapter that the phrase *bahirjyotih* refers to physical light with the help of which objects in the external world are visually perceived. *Antarjyotih* means " internal light " and refers to light of the subtler planes in which the temporary personality functions. Since the mechanism of perception on all these three planes is similar, namely, through the agency of a sense-organ and objects are seen by the help of vibrations coming from outside we may reasonably presume that lights of these planes are similar in nature, i.e., they all consist of electromagnetic vibrations but differ in wavelengths. According to Science the scale of electromagnetic vibrations extends on both sides to an unlimited extent and the lights of the subtler planes may therefore be considered as occupying different places on this infinite scale. The important point to note here is that all these lights according to the *śloka* are expressions of the Light of Reality.

How consciousness which is of the nature of Spirit can assume the form of vibrations which are of the nature of matter has been explained in another context. We need not therefore enter into this question here again. It will be enough to recall the general principle that the Supreme Consciousness being for all practical purposes the One Reality outside which and beside which nothing can exist, the only change which can take place in it must necessarily be of the nature of motion in its substance and this motion in its highest and subtlest form must be capable of assuming infinite number of forms, and through these forms providing not only the material basis of the manifested universe, but also the instruments of perceiving that objective universe in its different degrees of subtlety. Light, in its most general sense, belongs to the category of these instruments of perception on the different planes, and on the physical plane has been found to be an electromagnetic vibration which Science has investigated.

In spite of the enormous amount of research work which has been done in this field, scientists have been unable to unravel the mystery of the essential *nature* of light. In fact, the deeper they go into the problem and the more facts they discover about light the more confused they become. The situation has become so bad that they have given up the hope of unravelling this mystery and decided to remain contented with the empirical facts as they are, even though these are contradictory, utterly incomprehensible to the human mind and fail in the face of all common sense. In fact, they have thrown overboard common sense as a criterion of judging scientific theories and reserved it only for attacking all ideas which are not in accord with their preconceived notions.

The frustration which has overtaken the scientific world in its search for a Godless reality, can be judged to some extent from the following quotation from *The Universe and Dr. Einstein* (p. 35).

"Most modern physicists, however, consider it rather naive to speculate about the true nature of anything. They are 'positivists'—or 'logical empericists'—who contend that a scientist can do no more than report his observations. And so if he performs two experiments with different instruments and one seems to reveal that light is made up of particles and the other that light is made up of waves, he must accept both results, regarding them not as contradictory but as complementary. By itself neither concept suffices to explain light, but together they do. Both are necessary to describe reality and it is meaningless to ask what is really true. For in the abstract lexicon of quantum physics there is no such word as 'reality'."

This view is repudiated by another school of physicists including Einstein as will be clear from the following extract:

"Yet there are present-day physicists to whom the void between Science and reality presents a challenge. Einstein has more than once expressed the hope that the statistical method of quantum physics would prove a temporary expedient. 'I cannot believe' he says, 'that God plays dice with the world.' He repudiates the positivist doctrine that Science can only report and correlate the results of observation. He believes in a universe of order and harmony. And he believes that questing man may yet attain a knowledge of physical reality."

It will be seen from the above that the Theory of Relativity, though still based on the materialistic conception of there being only a physical reality in existence, is a tremendous advance from the old orthodox and rigid conceptions of the 19th Century scientists and has brought Science many steps nearer to the Occult doctrines. It has tried to unify all material forces, into one basic force. It has shown the illusory or subjective nature of Time and Space and Mass. It has shown that Motion is the basis of physical phenomena. It has shown that matter is only condensed energy and so there is no real matter in the strict sense. It has shown unconsciously that the subjective element plays a predominant role in the investigation of phenomena. All these are steps in the direction of the Occult doctrine.

Since we have already dealt with the nature of Sound and Light in its different aspects in some detail let us now summarize again the relevant Occult and scientific facts concerning the role of these two principles in the mechanism of manifestation and thus integrate all these facts into a composite picture which gives us an idea of this mechanism at least in vague outline.

(1) The first point to note is that matter and energy are two aspects or forms of the same reality which provides the basis of the objective universe. The two are inter-convertible, the condensation of energy leading to the formation of matter and the dissolution of matter releasing energy in a free state.

(2) Both matter and energy can exist in many forms and can provide an infinite variety of phenomena of different degrees of subtlety.

(3) Light and Sound are two fundamental forms of energy and though on the lowest plane both manifest as vibrations and produce sensations they play fundamentally different roles in manifestation. Though it is difficult to define and enumerate their functions with our limitations and incomplete knowledge it may be said in a general way that in the production of forms the function of Sound is to control and regulate the building up and maintenance of organized forms, and the function of light is to provide the raw material of these forms. This may be understood to

some extent by taking the constitution of an atom. Every atom is like a solar system in which several electrons are revolving round a central nucleus. Electrons are nothing but light whose dynamism has been reduced and which has been caught in this well-known structure as a result of the field of force produced by Sound and the appearance of electro-magnetic forces which are undoubtedly working within the atom. A similar phenomenon takes place in a solar system but the forces involved there are of a different nature and are called Gravitation and Inertia in scientific terminology. But the underlying force which regulates and controls these forces is again Sound.

(4) It is interesting to note here that Einstein has tried to prove in his General Theory of Relativity that the forces of electromagnetism (which include all other kinds of forces like cohesion etc.) are essentially of the same nature as those of Gravitation and Inertia and can be expressed by the same kind of field equations. This almost corroborates the Occult doctrine that both kinds of forces though outwardly different have a common source—Sound.

(5) The above paragraphs indicate the functions of Sound and Light in the production of forms whether the form is that of an atom, a solar system or a human body. But both Sound and Light have another and equally important and fundamental function and this function is related to the perception of the objective world consisting of forms. The perceptive functions of Sound and Light are easy to understand as far as the physical world is concerned because both exist in the form of vibrations which stimulate their respective sense-organs and produce the sensations of hearing and seeing. But this is the lowest aspect of perception. The conditions in the subtler worlds which interpenetrate the physical, change

progressively and the modes of perception change *pari passu*. Both Sound and Light continue to play a fundamental though changing role in perception on these planes.

(6) But, as in the production of forms, the roles of Sound and Light in perception are different due to the difference in their essential nature. The function of Light is related to the cognition of forms or whatever are the objective realities of a particular plane of existence as has been indicated to some extent in discussing the Sanskrit *śloka* in the previous chapter. The function of Sound, on the other hand, is concerned with bringing about changes in the subjective states of consciousness. We need not consider this question here. It is really a part of the theory of *Laya Yoga* and *Mantra Yoga* and can be discussed in appropriate contexts.

(7) It is necessary to note here that the fundamental differences in the functions of Sound and Light are due to the fact that they are related to the Principles of Consciousness and Power respectively or *Śiva* and *Śakti Tattvas* as they are called in Hindu philosophy. It is for this reason that the functions of Light are connected with the objective aspect of manifestation and those of Sound with the subjective aspect. This distinction is clearly discernible both in the production of forms as well as in their perception. Any one who studies this problem carefully and succeeds in understanding it, will be fascinated by the manner in which the Occult doctrines fit in with our general experience, with experiences gained in Yogic practice, and also with the recent discoveries of Science as far as they go.

Since Sound and Light both exist in several forms of increasing subtlety and the corresponding forms at each level may be considered to be related to each other this progressive subtilization may be represented symbolically by the following diagram showing the two ladders stretching from the lowest to the highest level:

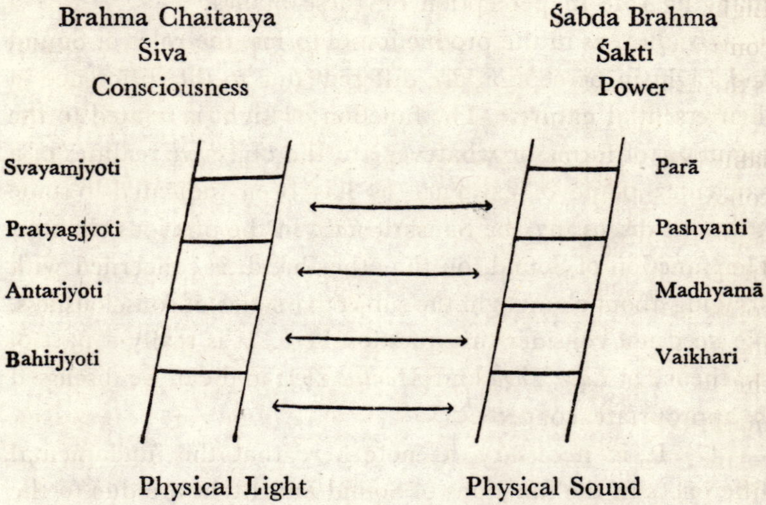

These four states may be considered to correspond with the four states of consciousness but as the matter has been discussed thoroughly in other contexts it is not necessary to go into it further.

As both Sound and Light are Cosmic phenomena we may also note here their relation to the five Cosmic Principles referred to as the *Pancha Bhūtas* or the Five Elements. According to the Occult doctrine *Ākāsha* is the primary Cosmic Principle from which all the other four Principles are derived and the order of derivation is given below:

Ākāsha ⟶ *Vāyu* ⟶ *Tejas* ⟶ *Jala* ⟶ *Prithivi*
(Space) (Vitality) (Radiation) (Fluid Matter) (Firm matter)

Since *Śabda* or Sound is the property of *Ākāśa* and *Prakāśa* or Light is the property of *Tejas*, it follows that Light occupies a lower stage than Sound in the chain of derivation. This means that the role of Sound is more fundamental than that of Light. This is in accordance with the fact that it is the controlling and regulating force while Light is related to and is the instrument of the objective mechanism and its perception.

(8) Since Sound and Light are two expressions of the same Reality which is dual and polar in nature each must contain both the principles. In Sound the positive principle is predominant while the negative principle is subordinate. In Light the opposite holds true. This conclusion is corroborated by the scientific fact that physical light is an electromagnetic phenomenon in which the electric element represents the negative and kinetic principle related to Power while the magnetic element represents the positive static principle related to Consciousness.

(9) The dual nature of Sound is shown by the fact that in its lowest form it produces repercussions on matter which are of the nature of vibrations which we recognize as physical sound. Its controlling and regulating function which is a static function is shown in a rather crude way by the formation of Chladni's figures when some sand is scattered on the surface of a drum and then a bow is drawn across the edge of the drum. Every kind of vibration produces its own pattern and even a slight change in the nature of the vibration is reflected in the modification of the pattern. This is a very simple illustration of the formative function of sound at the physical level but it shows at least that this capacity is inherent in it.

(10) Another interesting development has taken place very recently in the field of physics which has further corroborated the Occult doctrine with regard to the nature and function of Sound in manifestation. As has been pointed out

above all static forces like gravitation, magnetism, etc. are derived from Sound and though predominantly static should also have a dynamic character, i.e., should be able to give rise to waves on account of their dual nature. Scientists did not dream of associating waves with gravitation but Einstein predicted on the basis of his Theory of Relativity that there should be gravitational waves which may be difficult to detect. For more than fifty years doubts regarding the existence of gravitational waves persisted and it was only in 1970 that Einstein was shown to be right, after all, on the basis of experimental proof provided by Prof. J. Weber of the University of Maryland, U. S. A.

(11) It is this dual character of sound which enables it to serve as a bridge between different states of mind and Consciousness in *Laya Yoga* and *Mantra Yoga*. This is a general principle which we should keep in mind. Whenever two principles of an entirely different nature have to react, a third principle which combines the nature of both, even though in an artificial manner, is needed to serve as a *via media*. *Prāṇa*, *Kuṇḍalini* are all such intermediary compounds. Sound belongs to the same category when used as an instrument of affecting states of mind and consciousness. Its capacity in this respect is further increased by the fact that it simultaneously affects the vehicles of consciousness and attunes them gradually to the conditions needed for expressing different states of mind and Consciousness. These two effects operate simultaneously and reinforce each other. What has been said about the dual functions of Sound is also true in the case of Light, though in a different manner.

CHAPTER XXI

NEGATIVE WORLDS

THAT there exist negative worlds in which everything as we know it exists mysteriously in an opposite manner has been known to Occultism for a very long time, although for some reason the subject has been kept in the background and not openly discussed. Nobody took this Occult doctrine seriously and it was considered merely as a hypothetical assumption which Occultists and ancient philosophers adopted to make their philosophy subtle and thus to mystify the student. But now that Science has come to the conclusion that there must be negative worlds composed of the anti-matter which they have recently discovered, everybody has begun to talk of negative worlds and to wonder where they are and what they can be like. Their existence has been taken for granted by scientists because certain phenomena discovered concerning the nature of sub-atomic particles and the mathematical conclusions based upon them, point in this direction.

The existence of negative worlds is not a mere hypothetical assumption or flight of fancy from the Occult standpoint. In the first place, the existence of such worlds follows from the existence, at the heart of the universe, of an Ultimate Principle which is perfectly harmonized, integrated and whole, the Principle which we generally refer to as the Absolute. Since whatever exists in the realm of the manifest or the unmanifest is not only derived from this Ultimate Principle but also exists in a harmonized form in this Ultimate Principle, anything which happens within the realm of such a Reality must tend to disturb the harmony and distort the perfection of the Reality. It must therefore be balanced by an exactly similar thing which is just the opposite, so that the two together re-establish the equilibrium and maintain the void condition.

Everything in manifestation must disturb the harmony and distort the perfection of the Whole. So everything must have its equal and opposite in some form and somewhere. Time, space, matter, radiation and other constituents of the mechanism of manifestation must exist in two forms which are equal and opposite. And since these two opposite forms cannot co-exist in the same world, as we have seen in the case of sub-atomic particles and their anti-forms, there must be two kinds of worlds which are quite independent and out of touch with each other. One of these worlds must be composed of principles as they exist in our world, and as we know them, and the other world, which we may call the anti-world, must be composed of all principles in their anti-form.

In the second place, the existence of these worlds is known to the highly advanced Adepts of Occultism and they can function in them when it is necessary to do so. This is possible because these worlds are open to consciousness which can function without any instrument. The limitations arising out of the incompatibility of the two kinds of worlds apply only to the form side of manifestation and the instruments which function in these forms. Consciousness is above the realm of manifestation and so above the limitations inherent in manifestation.

Those who are interested in understanding the ultimate mysteries of existence, and particularly the mystery of duality which pervades the universe, should be clear in their mind with regard to the distinction between the transcendent and opposite states of a thing. There is a fundamental difference between the two although they appear the same to those whose ideas on the subject are confused. The opposite of a thing is its equal, on the same level, belonging to the same category, part of a *dvandva* or a pair of opposites. The transcendent state of a thing, on the other hand, transcends the pair of the corresponding equal and opposite things which merge and disappear in the transcendent state. It is at a higher level and is reached when the equal and opposite things are transcended.

This distinction between the two states is indicated by the prefixes, अ and नि in Sanskrit, नि generally indicating the absence and अ the opposite of a thing. But there is no sharp line of distinction between the significance of these prefixes, they are being used rather synonymously in many contexts.

The distinction between these two fundamentally different states is shown very clearly by the phenomenon of light. Light is white or coloured, white light being the integrated form of coloured lights. Both are positive things, a vibration, integrated or differentiated. Darkness or blankness on the other hand is the absence of light altogether, white or coloured. It is not the opposite of white light or coloured lights although it is generally considered as such. Anything when it is combined with its opposite destroys it or makes it disappear. Blackness or absence of light cannot destroy white light because it has no real existence on the plane in which white light exists. White light transcends coloured lights but darkness or absence of light transcends even white light.

Then, what is the opposite of white light? Obviously, it will be a kind of light which, when combined with white light, will make both disappear as light. Let us call it X-light. The clue to the relation of white light and X-light can be found in the relation of fundamental particles of the sub-atomic world and their corresponding anti-particles. A pair of such opposite particles, like the electron and the positron, when they combine annihilate each other and disappear into an equivalent quantity of radiation. They exist no more as particles. So this X-light which is the opposite of white light as we know it, when combined with white light, will make both disappear as light. Let us therefore call it anti-light for the present. This anti-light is thus the analogue of anti-matter in the realm of radiation. It should be the basis of the light phenomena in the anti-worlds which will be composed of anti-particles that have been discovered by Science.

Since the experimental discovery of these anti-particles by Science has made scientists consider seriously the existence of negative worlds of anti-matter and partially corroborated the Occult doctrine concerning the existence of such worlds, let us review briefly the startling developments which have taken place in physics in this direction during the last fifty years.

Every schoolboy knows these days that the structure of the atom is like that of a solar system, with a massive nucleus at the centre and a varying number of electrons revolving round it in orbits. Almost the whole mass of the atom is concentrated in its nucleus which carries a positive charge and is built up of varying number of protons and neutrons. The proton has almost the same mass as an atom of hydrogen which is the lightest among the atoms. The electron, on the other hand is a negatively charged particle with negligible weight which is $\frac{1}{1800}$th that of the proton.

Formerly, the protons and electrons were considered to be the sole constituents of all kinds of atoms, but subsequently, more and more new particles were discovered and the number of such particles, called fundamental particles, is now in the neighbourhood of 100. Of these, the neutron which is exactly like the proton but containing no electric charge, and the positron which is exactly like the electron but with a positive charge, are the most important.

The positron is called the anti-particle of the electron because when the two come together, they are both completely annihilated and an equivalent amount of radiant energy whose magnitude can be calculated from Einstein's equation $E=mc^2$ is given out. It has also been possible to create experimentally in the laboratory a positron-electron pair of particles from an equivalent amount of energy. So here we have a definite proof of the possibility of matter being transformed completely into energy on the one hand and material particles materializing out of immaterial energy on the other. This is in complete accord with the ancient Occult doctrine that the apparently material universe is produced from immaterial energy and is merely a play of *Sakti*, the Divine Power which can assume innumerable forms of energy.

It is interesting to note that Dirac had predicted theoretically in 1928 the above facts from mathematical considerations alone. He had predicted not only the annihilation and creation of the positron-electron pair but also of other pairs of particles like the proton-anti-proton or the neutron-anti-neutron. These pairs of particles also can now be produced in the laboratory by atomic bombardment. Other pairs of fundamental anti-particles like the meson-antimeson, neutrino-antineutrino have also been reported. Recent studies of these anti-particles indicate that there can be an anti-particle for each particle we know.

The prediction of the discovery of these pairs of particles from mathematical analysis by Dirac is also interesting from the philosophical point of view which has been discussed in dealing with the relation of manifestation and mathematics. If manifestation and mathematics are related to each other as a building and its plan in blueprint, it should be possible to predict such discoveries by mathematical analysis in this way.

After the discovery of these fundamental anti-particles the question arose whether these anti-particles could be combined and conventional atoms of the anti-variety could be synthesized. This work was accomplished for the first time by Dr. Lean Lederman in 1965. He was able to combine an anti-proton and an anti-neutron to create a nucleus of anti-deuteron or anti variety of heavy hydrogen. With the creation of this anti-atom it has been proved not only that both creation of matter from energy and complete conversion of matter into energy is possible but that the existence of anti-matter can no longer be questioned.

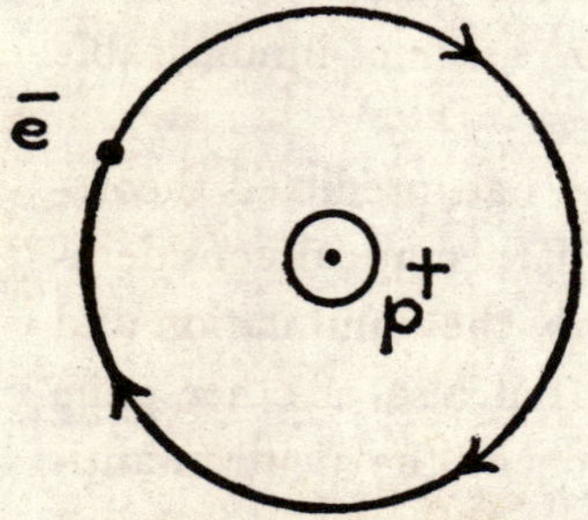

NORMAL HYDROGEN ATOM

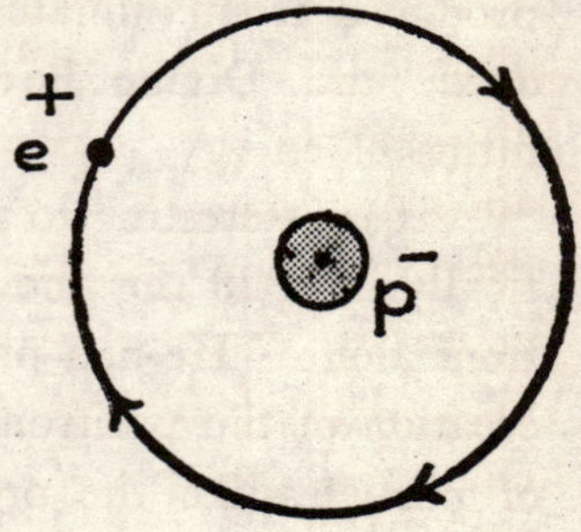

ANTI-HYDROGEN ATOM

In order to understand the relation of an atom and its anti-form, we may take first the example of hydrogen atom, the simplest of all atoms. This atom has only one proton in the nucleus and one electron orbiting the nucleus. In the anti-hydrogen, we shall therefore have one anti-proton in the nucleus and one positron or anti-electron orbiting the nucleus. These facts may be represented diagrammatically on the opposite page.

A helium atom has two protons and two neutrons in its nucleus. The anti-helium atom should therefore have two anti-protons and two anti-neutrons in the nucleus and two anti-electrons orbiting the nucleus. In anti-oxygen the nucleus will contain 8 anti-protons and 8 anti-neutrons and 8 positrons orbiting the nucleus. In the case of anti-carbon, there will be 6 anti-protons and 6 anti-neutrons in the nucleus and 6 positrons orbiting the nucleus. In this way, it is possible to visualize the structure of the anti-forms of all the ninety-two elements in the Periodic Table.

As all things like water, earth, plants, the bodies of animals and human beings and all physical objects are made from different combinations of these 92 ordinary elements, we can visualize anti-atoms of these ninety-two elements to combine similarly to form water, plants, etc., of an anti-world like ours with the only difference that all its atoms are of the anti-variety. We are thus quite justified in speculating about the existence of an anti-world which is like our own except that it consists entirely of anti-atoms and anti-molecules. A man of this anti-world though built entirely of anti-matter should behave in the same way, but if by any chance he comes in contact with an ordinary man, there will be an immediate explosion leaving a residue of matter or anti-matter from the body of the heavier man.

Matter and anti-matter are so anti or opposite in character that the slightest contact between them will instantaneously convert both of them into energy. This conversion will be accompanied by an explosion because conversion of matter into energy according to the equation $E=mc^2$ is cent per cent. Even in the most powerful explosion of the hydrogen bomb only about one per cent of the matter is converted into energy.

From what has been said above, it will be seen that there is no possibility of finding any anti-matter on our earth. It is also not possible to observe any anti-matter in space through a telescope because both matter and anti-matter are expected to emit the same kind of light. Then what is the basis for the supposition in scientific circles that anti-matter actually exists in some remote places in space?

There are some sources of tremendous amount of energy in space and it is impossible to account for these sources except on the basis of the interaction of matter and anti-matter taking place in these sources. The amount of energy which can be released by such inter-action can be judged from the fact that the mere interaction of one proton and one anti-proton—two sub-atomic particles—releases 1.8 billion electron-volts of energy. We cannot explain the continuing energy output of sources like quasars, etc., except on the basis of the theory that matter and anti-matter in large amounts is gradually coming in contact within them and releasing these tremendous amounts of energy by mutual annihilation.

Of course, this supposition merely pushes back one step the question where did all this energy come from originally? For, matter and anti-matter can be produced in equal amounts, as we have seen above, only from energy and such huge amounts of these two opposite kinds of matter could have been produced only from correspondingly huge amounts of energy according to the equation $E=mc^2$. Scientists are silent on this uncomfortable question and in their usual way ignore what they cannot explain.

It is a peculiar fact that when matter is produced from the materialization of energy equal amounts of matter and anti-matter are formed simultaneously. It has not been found possible to create an individual electron or any other individual fundamental particle out of energy, but only pairs of particles and anti-particles of all kinds. This fact is of great significance regarding the nature and origin of the universe and raises many interesting questions on which Occultism has to say a great deal and give very meaningful and definite information.

Science has not tried to understand the deeper significance and implication of this mysterious appearance of particles and anti-particles only in pairs and the impossibility of creating a single particle of anti-matter. This is not the place to discuss the philosophical implications of this mystery but it may be pointed out in a general way that it shows:

(1) Not only that matter can be created from energy but that there is a mysterious and infinite source of energy above the realm of so-called matter from which this stupendous material universe, requiring almost an infinite amount of energy for its production, has been created.

(2) This energy must be present in this source in a potentially dual but actually neutral form from which the pairs of opposite kinds of particles can be derived.

(3) This infinite source of energy cannot be material because matter, however stupendous in amount, cannot

be truly infinite. It is only mind and consciousness which can provide such a source. In the Occult conception of *Ākāśa* alone can we find a clue to the unravelling of this mystery. This question has been discussed in another context but the student of Occultism should ponder these problems deeply to be able to understand the significance of the remarkable discoveries made in this field of Science.

It has been pointed out above that the reaction between matter and anti-matter is considered as a probable source of the tremendous amount of energy coming from certain constellations or from the newly discovered quasars. This hypothesis does not, however, fit in with the current theories about the history and nature of the universe. If matter and anti-matter have been created in equal amounts, how did anti-matter get separated from matter before the two annihilated each other? If we assume that there is gravitational force of repulsion or anti-gravity between matter and anti-matter, then the General Theory of Relativity has to be completely discarded or drastically modified.

It has been suggested that the existence of anti-gravity can be tested simply by projecting horizontally a beam of anti-protons generated in an accelerator. If the beam is deflected upwards instead of falling downwards in the usual manner, this will prove the existence of anti-gravity or gravitational repulsion and render the General Theory of Relativity invalid.

The student will see from what has been stated in the last three paragraphs the fragile nature of all scientific theories which, like the Theory of Relativity, seem to be firmly established and are at present almost universally accepted. If the simple experiment suggested to test the existence of anti-gravity can be carried out and the beam of anti-protons in an accelerator is found to rise up due to gravitational repulsion the whole General Theory of Relativity will come down like a pack of cards and the physicists will again be found floundering in a sea of uncertain and mysterious physical facts. I do not say that this will happen but it can happen according to scientists and shows the uncertain and weak foundations upon which scientific theories are based.

In contrast to the uncertainty and ever-shifting nature of scientific theories the doctrines of Occultism which are based on the only certain and infallible method of direct perception have remained unaltered since times immemorial. If so-called scientific facts are found to disagree with these essential and authentic doctrines as known to the Adepts of Occultism it is the scientific facts or their significance which will have to change sooner or later and not the Occult doctrines. Scientific theories and philosophies have ultimately to fall in line inevitably with the Occult doctrines which represent the ultimate truths of existence discovered by the Adepts and verified again and again by every successful seeker after Truth.

सत्यमेव जयते नानृतम्

Satyameva jayete nānṛtaṃ

"Verily, it is truth which prevails ultimately and not falsehood or error."

We should not, however, confuse the results of clairvoyant research or mystic vision with the doctrines of Occultism. These former are also liable to error for reasons into which we need not enter here. So let not the sceptic throw such facts at the face of the occultist or demand proofs for the Occult doctrines.

CHAPTER XXII

THE PRINCIPLE OF UNCERTAINTY AND THE DIVINE WILL

LET us review briefly and in a general way the developments which have taken place in the field of Quantum mechanics and their impact on the Occult doctrine concerning the question of freewill and predetermination. According to Newtonian mechanics if we know the momentum and position of a body at one instant and the forces acting upon it we can predict both the momentum and position of that body for all future time with a precision only limited by our experimental skill in measuring the initial and attendant conditions. Newtonian mechanics gave a clear-cut and definite answer to the behaviour of a particle in motion, so much so, that Laplace made his famous statement: "If you tell me the present positions and velocities of all the particles in the universe and the forces acting upon these particles, I shall tell you all future history." Of course, it was realized that such calculations of future or past would for ever be impossible because of the innumerable number of particles in the universe but the significance of the principle lay in the fact that it meant rigid predetermination of the materialistic variety. It meant that the course of this manifested universe was predestined not by a capricious God of orthodox religions but predetermined by matter. Scientific materialism thus formally enthroned matter in place of God as the arbiter of our destiny and the future of the universe.

It is obvious that according to this principle all human actions are completely pre-determined and there is no place for freewill and much less for Divine Will in human affairs. Mind and Consciousness, it should be remembered, are considered as by-products of chemical, physical and biological reactions and have no independent existence of their own according to the philosophy of scientific materialism. This concept is embodied in another famous maxim of scientific materialism enunciated by Lombroso in his famous dictum: " The brain secretes thought as the liver secretes bile ".

The development of Quantum physics has revolutionized drastically the scientific concepts arising from Newtonian mechanics and created confusion and uncertainty in place of the clear-cut ideas and their dogmatic assertion which characterized scientific thought in the latter part of the last century. But before reviewing these developments very briefly let me quote a passage from Mr. Barnett's book which puts in a nutshell not only these developments but also their general impact on scientific thought and attitude. In chapter 4 of *The Universe and Dr. Einstein* he says:

" Quantum physics thus demolishes two pillars of the old science, causality and determination. For by dealing in terms of statistics and probabilities it abandons all idea that nature exhibits an inexorable sequence of cause and effect. And by its admission of margins of uncertainty it yields up the ancient hope that science, given the present state and velocity of every material body in the universe, can forecast the history of the universe for all time. One by-product of this surrender is a new argument for the existence of free-will. For if physical events are indeterminate and the future is unpredictable, then perhaps the unknown quantity called " mind " may yet guide man's destiny among the infinite uncertainties of a capricious universe. But the notion invades a realm of thought with which the physicist is not concerned."

The reader should note how relentlessly scientific developments are driving the modern scientists towards the well-established and time-honoured doctrines of Occultism and how desperately they are trying to evade the real issues by taking refuge in self-imposed and arbitrary ideals and objectives of Science. When their partial and limited scientific knowledge pointed towards the existence of rigid predetermination based on matter they proclaimed aggressively and with unconcealed glee that mind, freewill and their origin, God, do not exist. Now that most recent developments in Science have shown that matter in the scientific sense does not exist and material rigid predetermination is dead the vital questions concerning mind and freewill become 'a realm of thought with which the physicist is not concerned'. 'Verily, there are none so blind as those who *will* not see.' But how long can they fight this losing battle against the concept of a mental and spiritual basis of the universe proclaimed unequivocally and emphatically by Occultism? Even Nature seems to be against them and is bringing up more and more embarrassing and undeniable facts to their notice.

Let us now consider particularly an important development in the realm of physics, which has completely upset many well-established ideas of the scientific world and exploded the concept of rigid determinism upon which scientific materialism was based. This is called the principle of uncertainty.

It is not easy for the layman to grasp the principle of uncertainty in all its aspects, but the main and basic idea behind it can be understood to some extent if he is familiar in a general way with the latest developments in physics and chemistry. In explaining this principle in the simplest terms, let us recall what has been said above regarding the position and velocities of all particles in the universe, determining completely the past and future of the universe. It is this idea which provided the basis of rigid determinism and was considered as a pillar of scientific materialism.

According to the famous principle of uncertainty enunciated by Hiesenberg, Nature itself has put a limit on the precision with which we can measure both the position and momentum of a particle simultaneously. If we try to measure the position of a particle with greater accuracy, this leads automatically to the decrease in the accuracy of measurement of its momentum and vice versa. In other words, if we reduce the uncertainty in one, this automatically leads to increasing the uncertainty in the measurement of the other and the product of the two uncertainties can never be reduced below a minimum value which is equal to Planck's Constant.

The uncertainty introduced in our data in this way is not due to imperfections of our methods of measurement or those of our instruments. It is inherent in the very nature of things and we cannot do anything about it even if we are able to devise *ideal* instruments and methods of measurement: the defect will remain because it is built in, as it were, in the natural process itself and cannot be eliminated.

This peculiar behaviour of objects in the sub-atomic Quantum world is due to the fact that the act of measurement alters the very object being measured. How observation or measurement affects the object can best be understood by taking a simple concrete example. Let us suppose we want to measure the temperature of a cc. of water by means of a mercury-in-glass thermometer. When we introduce the thermometer into the water, the thermometer itself will take some heat from the water and the temperature shown will thus be considerably less than the actual temperature of the water. We can reduce the error by using devices with lesser heat capacity, such as miniature thermo-couples, but the error of measurement can never be eliminated completely. It is built in the very process of taking the temperature.

Divorced from its scientific terminology and experimental basis the principle of uncertainty really means that at its ultimate basis in the sub-atomic world the physical world does not follow the ordinary laws of the physical world and melts away not only from the consciousness of man but also from his grasp. The scientist can control the behaviour of these particles in mass but not of a single particle. He calls this uncertainty and lack of control over the individual atom 'the element of caprice in atomic behaviour' to avoid bringing in mind, consciousness and all other realities of existence which are hidden behind the physical world.

It is rather strange for the modern scientist to attribute this uncertainty in atomic behaviour to caprice when the whole philosophy and technique of Science is based on the tacit assumption that all processes of Nature take place according to natural laws which work with mathematical exactitude and can be relied upon under all circumstances. When a thing seems to behave in a capricious or erratic manner he always attributes this uncertainty to ignorance of some factors which are involved in the process under consideration and proceeds to find these factors to eliminate the uncertainty. But in all such cases where he suspects that superphysical factors may be involved and may force him to give up his materialistic philosophy his scientific outlook evaporates and he is prepared to throw overboard, logic, common sense, rationality and all other attributes of the scientific outlook. He begins to beat a retreat, refuses even to consider the problem further and takes a refuge in all kinds of excuses for not facing the realities which are involved in the problem.

The Occult explanation of this uncertainty present at the ultimate basis of the sub-atomic world is quite clear. According to the Occult doctrine this universe, in all its degrees of subtlety, is derived from and is the expression of an Ultimate Reality which pervades, controls and guides it from within. It can control and guide it only if the natural processes which underlie the universe are ultimately under its control. It is only then that its orderly progress towards its destined end according to a Divine Plan can be assured and brought about. In a physical universe governed completely by blind forces of Nature, which rigid determinism implies, there would be no possibility of such control and guidance and the universe will go relentlessly wherever such blind forces will take it.

The fact that this uncertainty is found only at the ultimate basis of the physical world, i.e. in the atomic world, is of great significance. It allows the universe in all its parts to run its course according to the laws of Nature which work with mathematical exactitude and can be relied upon under all circumstances. But when any change becomes necessary it can be brought about from within by working upon the individual atoms which are embedded in Divine Consciousness and can therefore be controlled by the Divine Will. Anyone can see that the Divine Consciousness which can control the individual atoms at the base of the physical world can control all physical processes which are built upon this foundation. So the simultaneous existence of determinism in the superstructure and uncertainty in the foundation of the universe allows the universe to run its course normally according to natural laws and yet its processes to be corrected by Divine Wisdom when this becomes necessary under any circumstances.

The manner in which the modern physicists have tried to avoid the difficulties arising out of the principle of uncertainty is ingenious. They maintain that "it is pointless for a physicist to worry about the properties and behaviour of a single electron; in the laboratory he works with beams or showers of electrons, each containing billions of individual particles (or waves); he is concerned therefore only with mass behaviour, governed by the laws of probability and chance. So it makes no practical difference whether individual electrons are particles or systems of waves—in aggregate they can be pictured either way."

This conclusion would be valid only if there is only the physical world of atoms and molecules in existence and the individual atom or electron is just a particle of insentient matter with nothing else in its background. Then its behaviour would be governed solely by the laws of probability and predicted on the basis of the laws of Nature. But if each individual ultimate particle is embedded in the Divine Mind and Consciousness and amenable to Divine control then the laws of probability will govern its behaviour only in normal conditions. It will also be possible for the underlying Mind and Consciousness to alter the behaviour of the particles *en masse* and achieve any purpose through them. It is only mind and consciousness which can take initiative that can alter the behaviour of particles *en masse* in this manner. Mere insentient particles cannot do this.

To put it in a rather crude language God has kept the initiative and the power to control the universe according to

His Will in his own hands and not left it either to insentient atoms or to unscrupulous or undeveloped human beings who would like to mould it according to their selfish purposes or immature judgments. The element of the unpredictable and unexpected is inherent in the very nature of the manifested universe and can be seen even in human affairs by those who have the eyes to see these subtler things and whose minds are not completely immersed in its phenomena. Man is free to do what he likes but God is also free to make him bear the consequences of his actions either through the ordinary operation of *karmic* laws or through extraordinary circumstances which He can create for this purpose.

What has been said above with regard to the supremacy of Divine Will in reference to physical matter holds true equally with regard to mind. Every individual mind is embedded in Divine Consciousness and is therefore amenable to the control of the Divine Will. Mind being more pliable than physical matter should be easier to control but perhaps these differences do not matter where infinite Will and Power of the Divine Consciousness are involved. The lives of human beings as individuals or communities are allowed to run their normal course according to the physical and superphysical laws applicable to them. But where extraordinary circumstances arise, Divine Consciousness can intervene to put things right. Nothing can interfere in this because this Consciousness works from behind the mind and the mind is not even aware of this direction from within.

